Written, edited and designed by the editorial staff of ORTHO BOOKS.

Coordinating Editor:
James K. McNair

Manuscript and Project Designs:
Timothy Downey

Special Consultants:
Clyde Childress
Alice H. Quiros

Photography:
William Aplin
Ernest Braun
Clyde Childress
Tyler Childress

The Facts of Light
about indoor gardening

Contents

More light on indoor gardens

Plants indoors and their relationship to light. Good design for interior plantings—from a window garden to an entire house.

This book is about *plants in the house,* not about house plants. Bookstores have whole sections on house plants, with many excellent sources covering both general and specialized indoor gardening. Nor is this book specifically about Light Gardening, the hobby that has brought pleasure to thousands. The emphasis here will be how to garden indoors, zeroing in on solving lighting problems, general cultural needs and how to make plants an integral part of the home through careful design.

Plants belong in every room (there's even a reason to put poinsettias in the closet!) and there are comparatively few real restrictions. Bringing light to plants and plants to light—this book tells you how to do that.

The homeowner must recognize that plants indoors are in an artificial environment—one considerably different than their native habitat. Few living rooms have the Mediterranean climate that has sent us the Jerusalem cherry, the tropical climate of philodendrons, or the desert environment which nurtures cacti and other succulents. The artificial atmosphere in which plants indoors are asked to survive, if not thrive, differs from their native habitats in significant ways:

✔home environments tend to have relatively constant temperature and low humidity;

✔pots restrict root growth and the soil they contain fluctuates in fertility and moisture;

✔there is little or no air movement or rain to refresh plant foliage;

✔watering, unlike natural rainfall, is often of poor quality and usually

Patrons of Lehr's Greenhouse Restaurant, San Francisco, dine in a tropical garden.

too much or too little;

✔feeding tends to be concentrated;

BUT PARTICULARLY:

✔light is generally from one direction;

✔darkness is of a shorter duration (assimilation of nutrients takes place during dark hours);

AND EMPHATICALLY:

✔*there is usually much less light in the normal indoor situation.*

IN THESE PAGES WE WILL EXPLORE:

✔the general nature of light;

✔an understanding of proper plant exposures;

✔how to make the best use of sunlight by proper exposure and intensity; and the use of "windows" to make this possible;

✔modification of light by curtains, mirrors and other techniques;

✔what to do if there isn't enough light to keep plants healthy;

✔remodeling to direct more light to the proper places.

Hopefully, the ideas described here will stimulate the readers' imagina-

Outdoor blooming shrubs are enjoyed indoors in the Douglas Jones' two-story living room.

Outdoor tree branches filter early morning light falling on plants in garden writer Maggie Baylis' second-story breakfast room.

tion to carry indoor gardening expertise one step farther. Every environment is different and with a little imagination almost any lighting problem can be solved.

Estimating the amount of light available for plant growth is absolutely essential in order to select the right plants and place them in the proper setting. Several techniques will be described to measure light levels or to approximate light intensity in the home.

We'll name names—both common and Latin—of house plants with their lighting requirements necessary for their healthy growth. Plants that can take direct sunlight will be pointed out, as well as those that can survive in usually dim northern light. Long-day and short-day plants will be identified along with the special care involved in making them perform.

Artificial light is man-made sun—it can make a tropical garden bloom in an area devoid of natural light deep in the canyons of Manhattan. It can be used as a supplement to natural sunlight or as a sole light source.

The seasons can be anticipated with indoor projects: Summer vegetable and flower gardens can be started indoors during winter and early spring. Seedlings can be grown stronger and faster under artificial lights.

Included are projects the homeowner can undertake to design with light, making indoor gardening not only a successful and enjoyable experience, but an indispensable part of the home environment.

Indoor gardeners are often baffled by so much available information. The following excerpts from an editorial by Dr. H. M. Cathey, in *American Association of Nurserymen,* may help clarify some of the conflicting advice you receive.

"If you assemble more than one gardener in the room and begin discussing the care, feeding and sunlight requirements of any species of plant, you'll soon find an argument in the brewing. Even the environmental requirements of a simple plant like a coleus may start a raging controversy. Although most gardeners will agree that filtered sun is best, very few individuals can come to an agreement about the proper fertilizer and mois-

Skylights in restored Victorian flat were designed by Timothy Downey for client Charles Camp. They incorporate a special ledge to hold container plants. Metal trays contain moist gravel and catch dripping water. In the evening (lower photo) built-in lights illuminate the plants.

Above: The specimen fiddle-leaf fig and other plants receive light from two directions. Below: Employees of Esprit De Corp., a clothing manufacturer, enjoy the plants in the factory; maintain them during "coffee-gardening" breaks.

Large windows provide light for Asparagus "fern" on balcony and plants below.

The Charles Foster Restaurant, Atlanta, offers relaxing dining among potted trees.

Opposite page, upper: Mrs. J. T. Austin, Atlanta, supplements sunlight with plant growth fluorescents installed on either side of the windows. Opposite, lower: Plants are combined with memorabilia in Stanley Paul's well-lighted hilltop kitchen. His bathroom (above) is a cool, humid retreat for plants and people, designed by H. Judd Wirz and Associates.

ture levels for these relatively hardy plants.

"Check any gardening book to examine our current state of knowledge. Read and compare the suggested frequency of the watering of a Norfolk Island pine or a Boston fern. Now check another source. You will note that one suggests that you water almost daily; others encourage you to train the plant and water it only once a week. Now, mix and stir these views with other monographs written by noted garden authorities and you find hundreds of opinions on syringing, the proper fertilizer, the correct sun exposure, the best way to control insects, proper pruning methods, propagation techniques . . . etc. The opinions are endless and, oddly enough, to a degree, may be correct.

"If we draw a conclusion from all these noted authorities, we discover that plants are extremely tolerant of a wide range of abuse from the cultural procedures to which we subject them . . . plants adjust their growth characteristics and survive often under the most difficult conditions. Yet many plants will die abruptly if too much abuse is given them. Others just slowly 'do the strip'—as one leaf after another falls from the plant.

"Most gardeners take their failures in stride. They simply try again . . . and again . . . and again, often reading every monograph they can find on the subject. Eventually they profit from their previous mistakes of not enough light, moisture or nutrients. Eternal optimism rules the mind of the good gardener."

The aim of this book is to profit from the good indoor gardening practices learned from the good indoor gardeners.

Shown on pages 8-11 is the ultimate in indoor gardening—a northern California redwood house designed by David Clayton for a couple who wanted a 'perfect' environment for their extensive plant collection. Many skylights and windows of every shape and size were incorporated into the design to capture as much natural light as possible.

Attention was given to details that make indoor gardening a little easier, including a greenhouse, potting shed entrance, terracotta floors that water can't damage, pulleys for lowering plants to water and special places for wet plants to drain.

Consider plants you'll want to grow when you design or remodel a space for living or working. Incorporate some of the good ideas found within these pages into your home when building, remodeling, or modifying.

Designed for indoor gardening: *Located in a quiet, wooded retreat overlooking the Pacific Ocean is a redwood house designed by David Clayton. Every day isn't as perfect as the one our photographer caught, so the owners wanted a space where they could grow plants in spite of the oft-prevailing oceanside fog. Thus the decision to open the house to capture as much available sunlight as possible.*

1. *Two-story living room window walls and skylights supported by huge re-cycled beams provide brief afternoon sun. Outdoor trees soon filter the light preventing indoor leaf burn. The entrance side of the house faces the east.*

2. *Architecture features skylights and many windows of all sizes and shapes. Guests enter the house through a combination greenhouse-potting shed, passing through a conservatory of specimen plants.*

3. *The greenhouse area, in front, to the left as you enter, houses plants that are undergoing special care, along with newly propagated species.*

4. *The conservatory, to the right of entry, has terracotta floors with drains, making plant watering an easier job.*

1. *Plants in the bathroom are grouped on a built-in ledge underneath a trapezoid-shaped window that follows the lines of the roof. This tends to tie the indoor and outdoor foliage together. The tropical plants enjoy additional humidity from running faucets and an adjoining steam room and Jacuzzi.*

2, 3 and 4. *Hanging plants are bathed in eastern sunlight from the top by small skylights. Extended windows below these allow ample light as the sun shifts through the day. All hanging pots in the house have temporary clamp-on plastic bowls to catch spills or drainage. These are made from plastic food storage containers with attached coat hanger wire hooks.*

5. *In this view, looking into the house from the outside, the eye is fooled into thinking sunlight is coming from inside. Actually, it's the afternoon sun, shining through the rear skylights, falling on the back of the plants.*

6. *This photo is taken from the living room looking into the conservatory area. It gives the impression that plants in the conservatory are out-of-doors. Through intentional design this "feeling" is carried throughout the house; changing as the day's light changes.*

The photography for these 4 pages was scheduled and completed in one day so that the staff and photographer could capture the lifestyle of this type of architecture. In spite of putting in a very full working day, they all agreed that the atmosphere of the house made the day seem almost relaxed.

4

5

6

rain bottle
SUDBURY
SOIL TEST KIT

Setting the stage

Evaluating your indoor garden area. How-to-build convenient work spaces. Plus a few basics for healthier plants.

Indoor gardeners are a diverse group ranging from the kindergarten enthusiast, who first sprouts a bean seed, to the retired professional who's mastered the art of orchid hybridization. Even persons who are not ambulatory can enjoy miniature gardening at their bedside.

You are an indoor gardener even if you only possess one plant—a living gift for a special occasion or a coleus cutting you've nurtured along. Your 'garden' may be the kitchen windowsill, a fluorescent-lighted bookcase, or even your place of employment where you combine gardening with your coffee break.

Indoor gardeners do not 'grow' plants—the plants grow themselves. Our responsibility is to provide the right environment in which a plant can flourish. Unlike outdoor plants which can send roots out in search of water or food, indoor plants are confined to containers and are dependent upon the gardener to supplement Mother Nature.

Even with all good care, we must be prepared to accept the inevitable. Like all living things plants eventually die and must be replaced.

Help for the Indoor Gardener

Everywhere we look today there are plant shops and boutiques, plant departments in major stores, nurseries, revitalized florists; even supermarkets and hardware stores have plant sales areas. Indoor plant and gardening supplies are big business as more and more people discover the excitement and challenge of indoor gardening.

To meet the demand,countless new products have hit the market. From this vast array of gadgets, accessories and devices, are quite a few helpful items for indoor gardening. The photo on left shows just a few examples of what is available, taken from our random sampling of retail outlets.

Analyze your own personal gardening needs, tastes and desires before visiting your local nursery, garden center, hardware store, plant shop, or department store. This will enable you to select only the items that you will find really useful in your personal garden.

Home Environmental Inventory

Unlike the house shown on pages 8-11, very few interiors offer ideal gardening facilities. Most homes, shops, or offices are too hot and dry for most plants and have too little or, in a few cases, too much sun.

Take a look at your space from the plant's point of view. First, consider the light. Can you meet the plant's light requirement with either natural or artificial means? Use the chart, pages 28 to 41, as a basic reference. Reputable nurseries and plant shops can also help you make an educated selection after you've selected a plant 'site'.

In addition to light you should take an inventory of humidity. Where are the best places to locate plants where they will receive needed moisture from the air? Kitchens and bathrooms with their steam and running water are obvious spots. But kitchens also have ovens and ranges giving off excessive heat and using up moisture in the air. This frequent heating and drying action can offset the advantages of running water. Place a thermometer where you wish to place any plants. Check it during the time your range is getting maximum use. Misting air can minimize effects of higher temperature by replacing moisture lost through heating. And don't forget ways to increase humidity in other rooms. Some plants, however, like cactus, prefer dry locations.

Another important, and often neglected, consideration in locating plants is the design aspect. Where will it look best? How does it relate to the architectural design and interior decor? Apply the same principals used in placing furniture or works of art when you plan your interior 'horticultural design.'

Interior spaces have many microclimates just like the backyard garden. There are hot spots, shady nooks, damaging drafts, along with perfect growing spots. Seek out those places where plants will thrive. Don't be afraid to try plants in many places. Of course, you must be prepared to live with the results. One gardener reported that her piggyback is only happy on top of the refrigerator and she has to lift its leaves whenever she wants to open the freezer door. But she feels that the reward of a beautiful, thriving plant is worth these extra efforts.

The Indoor Potting Shed

One of the major problems, when your house is your garden, is where to do all those simple, but messy, chores. You can't repot without getting into the soil and this can be a problem in the living room.

On the following pages several suggestions are offered for creating a workspace that will confine your garden chores to a convenient area.

A few samples of indoor gardening equipment and accessories currently on the market.

Kitchen "potting shed"

Incorporate useful ideas into your kitchen design to create an indoor gardening area. Storage is provided underneath the sink and in large glass jars on the counter top. Make a wooden rack to fit the inside top of your sink for draining plants that have just been watered. Hooks over the sink are great for hanging baskets while they drip. The lightweight hose can be taken into other rooms to make watering a little easier.

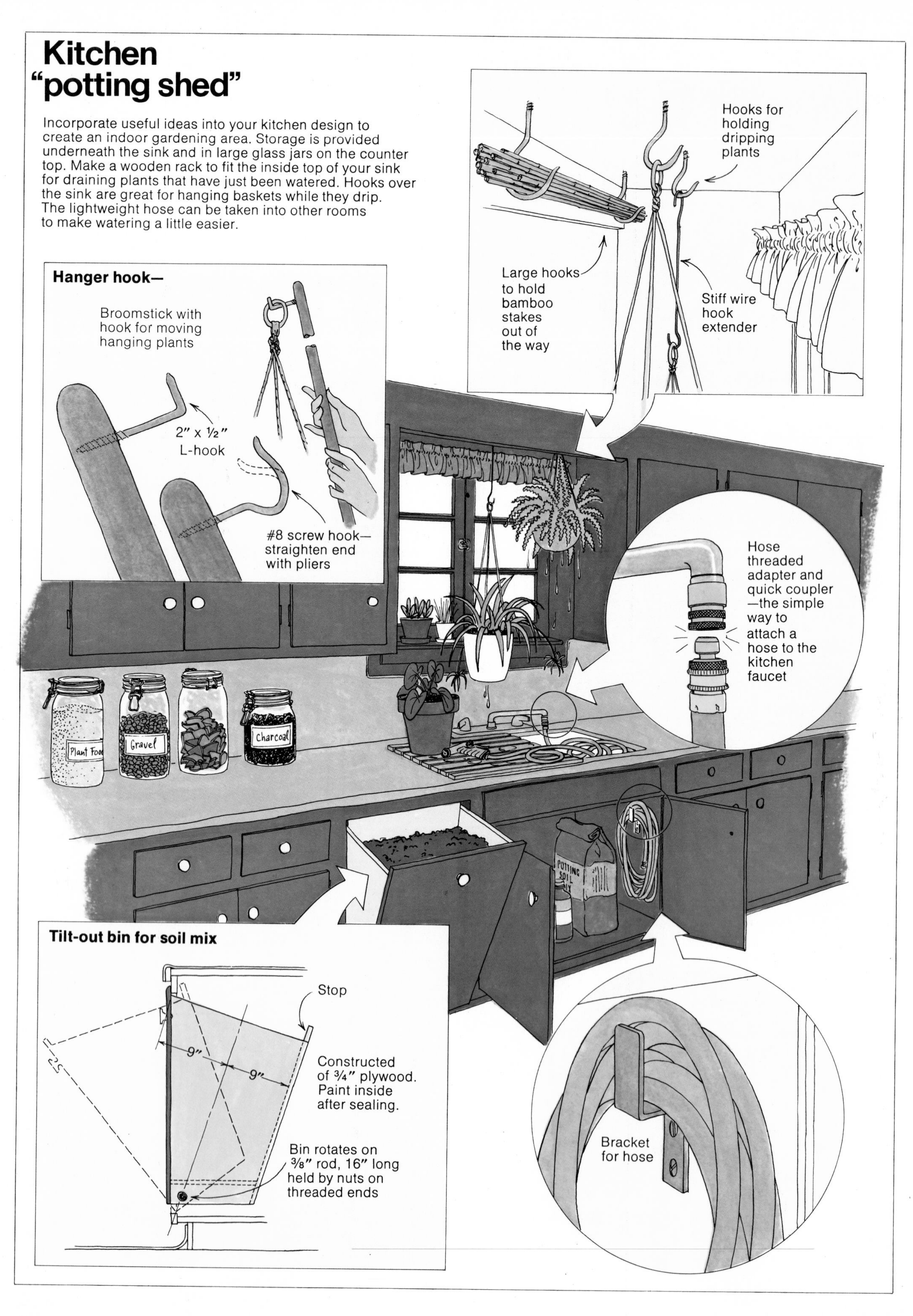

Closet garden workshop

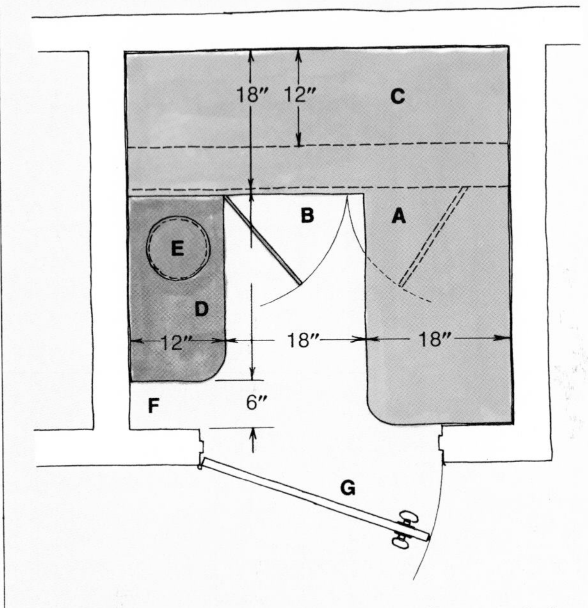

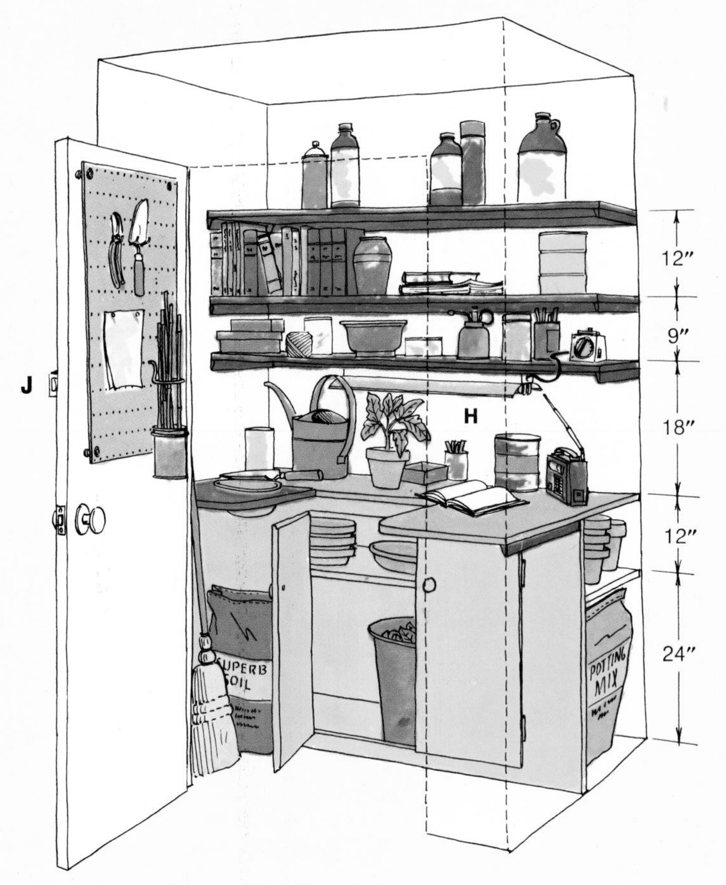

Here are two ideas for turning a spare closet—walk-in or standard—into an indoor gardening area. Both provide ample storage for tools, pots and garden products and offer a retreat for all those messy gardening chores that can be shut away from the rest of the house. Provide plastic drop cloths for catching spilled soil, water, or leaves. Plant growth lights on automatic timers allow you to leave problem plants inside during periods of recuperation. With careful planning you can utilize every inch of the area, even the back of the closet door.

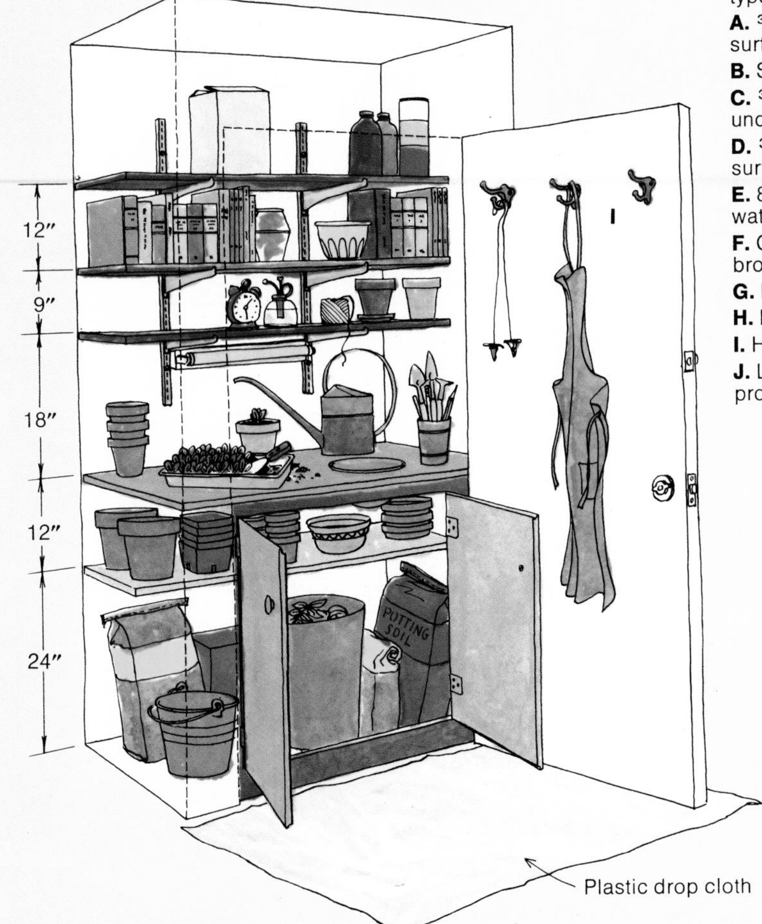

Key to floor plans of walk-in (above) and standard type closet (below):

A. ¾-inch plywood work counter with plasticized surface (e.g. 'Formica').

B. Storage cabinet underneath counter for storage.

C. ¾-inch plywood shelves over counter and in cabinet underneath counter for storage.

D. ¾-inch plywood work counter with oil base painted surface, butcher block, or plasticized surface.

E. 8-inch plastic bowl set into counter to catch excess water and soil.

F. Corner storage for long-handled tools such as mops, brooms and plant light on tripod (see page 95).

G. Pegboard with hooks for tools.

H. Plant growth lights on automatic timer.

I. Hooks for extra tools, hangers, aprons, etc.

J. Lock to keep children and pets away from garden products.

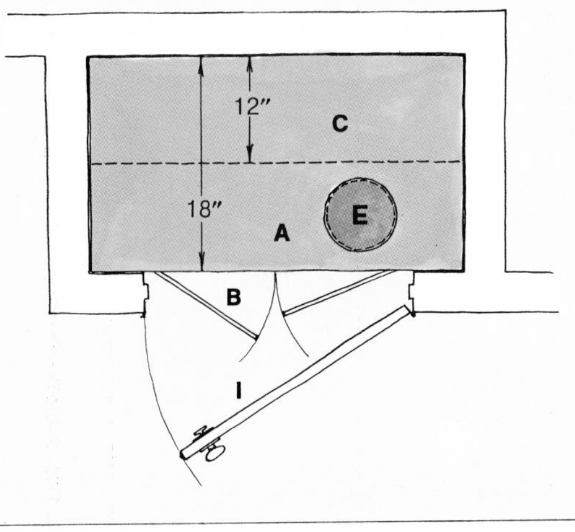

Plant cart and display cube

The simple lines of this attractive display unit fit into almost any decorating plan, depending on the finishing materials and colors selected.

The width is designed so the cart can be wheeled through almost any door for on-the-spot gardening. When you're ready for gardening chores, simply open the drop leaves for an expanded work surface. Inside the cabinet are shelves and drawers of all sizes to hold all the necessary tools and supplies.

The height of the cart allows it to double as a service bar or buffet.

Most of the construction is of ½-inch and ¾-inch birch plywood. Top, drop-leaves and sides are Formica or plastic covered. Other materials needed include:

4 pair shutter hinges
2 18-inch piano hinges
2 5¼-inch piano hinges
1 pair 18-inch drawer slides
3 pair 14-inch drawer slides
4 touch latches
4 casters
1 handle for guiding (optional)
Miscellaneous nails, glue, screws, sandpaper and paint.

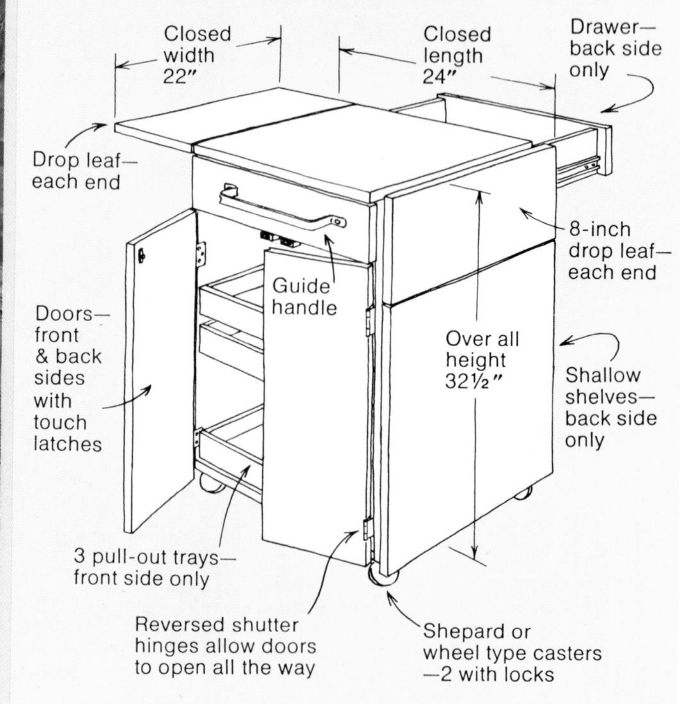

A few basics

Here are some simple guidelines for the indoor gardener.

Soil. The most reliable and easiest method is to use pre-packaged sterilized soilless mix such as Jiffy Mix, Pro-Mix and Redi-Earth. The universities have done the experimentation for you and have developed these soil mediums for growing all types of indoor plants. If you want to mix your own soilless growing medium, here is a recipe for Cornell mix:

4 quarts #2 grade vermiculite
4 quarts shredded or fine grade peat moss
1 level tablespoon superphosphate
2 tablespoons limestone
4 tablespoons dried cow manure or steamed bone meal.

This will yield enough mix to fill the equivalent of 7 to 8 six inch pots.

Some indoor gardeners have developed favorite soil mixes through the trial and error method. If you have time, space and the inclination, you can experiment with various proportions of sand, peat moss, leaf mold, vermiculite and ingredients such as ground limestone, crushed volcanic rock, tree fern bark, redwood chips and fertilizers, to create your own mix.

Soilless additives have various definite reasons for use, as follows:
Vermiculite: helps to retain moisture and slows compacting

Perlite **Crushed volcanic rock** **Pebbles** **Redwood chips** **Tree fern bark** **Charcoal**	**Lighten soil mix, increase drainage, prevent compacting**

A good all-purpose potting soil can be made by mixing together two parts sterilized (oven 180°—½ hr.) garden soil, one part leaf mold from the compost pile or packaged humus, and one part sand. This suits plants such as geraniums, amaryllis, dracaenas, palms and oxalis.

High humus content is important for such plants as African violets, begonias, philodendrons, etc. They do well in a mixture of equal parts sand, peat moss, leaf mold and a balanced garden soil.

Plants from the desert need a gritty, lean-growing medium. Most cacti and other succulents will prosper in a mixture of one part garden soil, one part sand (beach sand should not be used. Coarse "builders' sand" is type to use), one-half part decayed leaf mold and one-half part crushed clay flowerpot or brick. To each half bushel of mixture, add a cup each of ground horticultural limestone and bone meal.

Containers. Plants can grow in almost anything that will hold soil. The classic clay pot is hard to beat, both

An old table was converted into a combination garden work cart and artificial light unit. Wooden fruit crates hold soil mixes and tools. Wheels allow the unit to be easily moved.

aesthetically and from the growing standpoint; allowing the plant's roots to 'breathe' through the porous clay. However, there is evaporation with clay which means that the rootball may dry out quickly. This porous quality causes seepage, so you will need to protect tables and carpets with a rubberized, plastic or clear glass saucer, or a piece of cork cut to fit the underside of a clay saucer.

If you can't resist watering, then by all means choose clay. If you are on a busy schedule, you may prefer to use plastic. There are plastic pots made in a wide range of colors and sizes. One advantage of plastic is that you don't have to water as often as clay pots. They must be watered carefully to avoid overwatering. The test is when the soil feels moist but not sopping wet.

There are many types of glazed pots available. You can choose simple cylinders or elaborate handcrafted decorative pots. The most common practice is to place the plastic or clay pot inside the decorative one. (Be sure that inside pot is not standing in water drained into liner pot.) Or you may transplant directly into the decorative container. See directions to the right.

Wooden planters of redwood, cypress, or marine plywood are perfect for many large plants and trees. And there are baskets of every description to hang or sit (with plastic or metal liners inside to prevent soil and water from leaking through). Container choice is dictated only by your taste and imagination.

Repotting. Most plants do not require repotting nearly as often as people think. Many grow best when roots are a little crowded forcing top growth. The rule of thumb about repotting is to *move the plant up only one pot size at a time.* Don't jump from a four inch pot to a huge container all at once. Check drainage holes to see if any roots are growing through. In a pot without drainage holes, it's a good idea to turn the plant upside down occasionally and tap the pot all around, resting it on a sink and slipping the plant and soil from the pot. Check the rootball. If it has filled the container, with a solid mass of roots on the outside of the ball, then it's probably repotting time. If not, return it to the same pot and firm the soil.

If you are repotting in a clay pot it's advisable to soak the new pot in water for about 30 minutes so that the porous clay won't rob the roots of soil mosture. Cover the drainage hole of any container with a piece of curved broken crock or small stones. This

Planting in a decorative container

Here are three methods of planting in decorative containers with no drainage holes. The first simply inserts the existing pot into the container. Disguise the pot with moss, bark, or pebbles.

In method two the pot is placed inside the decorative container and stuffed with loose medium such as perlite or vermiculite to hold moisture. A little charcoal keeps water smelling sweet and appropriate coverings disguise the pot.

Method three has the plant actually planted in the container, with fine gravel on the bottom to catch drainage. The plastic tube and dipstick allow you to check excess water to help prevent overwatering.

In any of these methods it is important not to overwater!

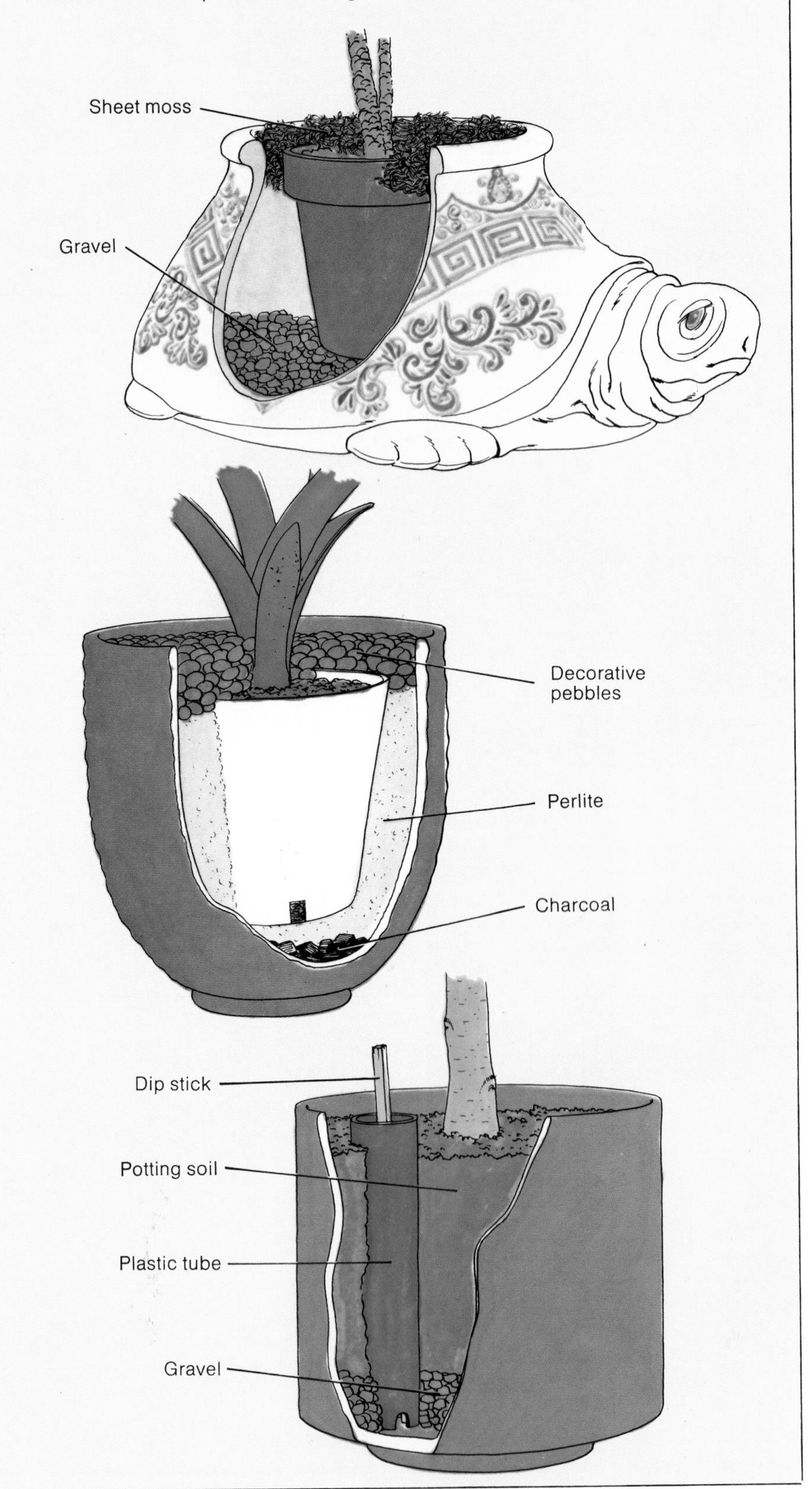

keeps soil from washing through, but allows water to drain. (If sand or desert grit is used, window screening will prevent soil loss.) Add fresh soil mix, place the rootball inside and fill all around with new soil. Set the plant at the same level that it was growing. Be sure to pack well, pushing in with fingers to prevent any air pockets. Lightly tap the pot on a hard surface to settle the soil. Water well (except cacti and other succulents, which should wait a week before getting water) and clean the pot of loose soil.

Should you decide to repot in a container without drainage holes, add an inch or two of charcoal chips, volcanic rock, some gravel, or broken crock for catching excess water. Then add some soil and continue as above. Be extremely careful not to overwater in these containers to prevent root damage.

Some people insert a section of plastic tubing down into the pot so they can check excess water with a dipstick. Overflow will evaporate or can be suctioned up with a syringe (turkey baster).

Watering. No one knowledgeable in plant care will tell you to give a plant one or two cups of water, or any other specific amount. The rule is to *water when the soil feels dry to your finger about ½ inch to an inch below the surface. Add tepid water until it drains through the bottom.* Frequently, after severe dehydration of the rootball, it is necessary to add a liquid detergent to the water (1 tsp. per gallon). This breaks the surface tension of the water and allows it to penetrate the soil. Submerging a pot in water to its rim, a procedure called bottom watering, takes more time, but it is good for plants you've allowed to dry out severely and for those you get in full bloom from a florist. Submerging in a pail of water is the best way to water plants, such as bromeliads or staghorn fern, growing on slabs of tree fern or fiber; as well as hanging baskets made of this material. Allow to drain, then return to growing space.

Whether you apply moisture from above or below, pot saucers need to have excess moisture poured off within an hour after watering. Never leave plant standing in water.

There are some plants, however, like maidenhair ferns and Cyperus which need more water and should be moist at all times. Others, like cacti and succulents, must dry out thoroughly between waterings. See chart on pages 28-41 for a specific plant's water requirements. Several new devices are marketed to indicate when a plant needs water—from sticks and papers that indicate color changes to sophisticated meters which can be inserted into the soil for a reading of soil moisture.

New automatic watering devices help simplify your chores and are great to use when you are away from home. Water wicks provide a constant water supply by drawing water from an adjacent container as the plant needs it. Some pots have built-in water reservoirs to draw from whenever the soil is dry.

Good news for container plants and hanging baskets—*Viterra* is a new product with the *Hydrogel* amendment mixed into the soil medium to increase the water holding capacity. It lets plants go longer between waterings and helps eliminate overwatering problems. It has been tested by commercial bedding plant growers. One report says "We have done enough work with Hydrogel this past year to be very excited about its possibilities. It should be something good for salespeople and consumers who do not water properly." Viterra is currently available in a potting soil sold at retail through the W. Atlee Burpee Company, Warminster, PA 18974 and is expected to be marketed by a number of other companies in the near future.

Most drinking water is safe for your plants; however, avoid using softened water which is high in sodium (most water softeners work only on the hot tap). Distilled or spring water is excellent but costly, and not normally necessary. How you use water is more vital to plant life than the chemical content. More plants are killed from too much water than from not enough.

Humidity. There are many ways of increasing humidity for plants—short of converting your house into a greenhouse. Try grouping plants together in a large tray filled with pebbles. Maintain a water level below the pebbles, avoiding contact with the pots. The tray also catches the excess after watering.

Small containers of fresh water can be placed among a plant collection. Extra humidity is gained as it evaporates.

In dry places, spray leaves (shiny or smooth foliage—not "hairy" leaves, nor fern fronds). This will also keep them clean. If plants are located where water mist can damage furnishings, spread a plastic dropcloth before misting. Or try enclosing the entire plant in a large plastic bag and misting through a hole in the side.

Portable humidifiers can be positioned in dry rooms, or an inexpensive humidifier added to a central heating system to add moisture to the air.

Automatic watering devices

Washing leaves with tepid water cleans off layers of dust, actually getting more light onto the surface, as well as helping get rid of pests before they become a problem.

Feeding. Container plants need regular feeding when they are in active growth. Many houseplant fertilizers on the market have been formulated for use every two weeks. This is more effective than large monthly doses. Follow container directions for rate and frequency.

Fertilizers are available in all types: water soluble pellets, powders and liquids; dry tablets and sticks to insert in the soil; and time-release pellets. Whatever kind you choose, read the label first and follow the directions. Make sure to moisten the soil before adding fertilizer to prevent root burn.

Grooming. Plants indoors need regular pruning and grooming to keep them attractively shaped and to a reasonable size. With soft-stemmed plants like Begonias, geraniums, Coleus and wandering Jew the growing tips should be pinched off to encourage branching. Plants with woody stems like Hibiscus, Gardenia and jasmine can be pruned with small hand pruners to keep them compact. After flowers fade, removing dead blossoms will keep the plant looking neat and will promote better growth.

Leggy, top-heavy plants or vines may need to be staked and tied for support or attached to some type of trellis or 'totem pole' on which it can attach itself.

Topsoil should be kept loose and clean. Coverings of moss, stones, or bark can be used, especially useful in camouflaging a plastic pot inside a decorative container. These act as a mulch, also, preventing rapid evaporation of moisture from the planter mix—especially useful in dry, hot conditions.

Tip pinching or pruning

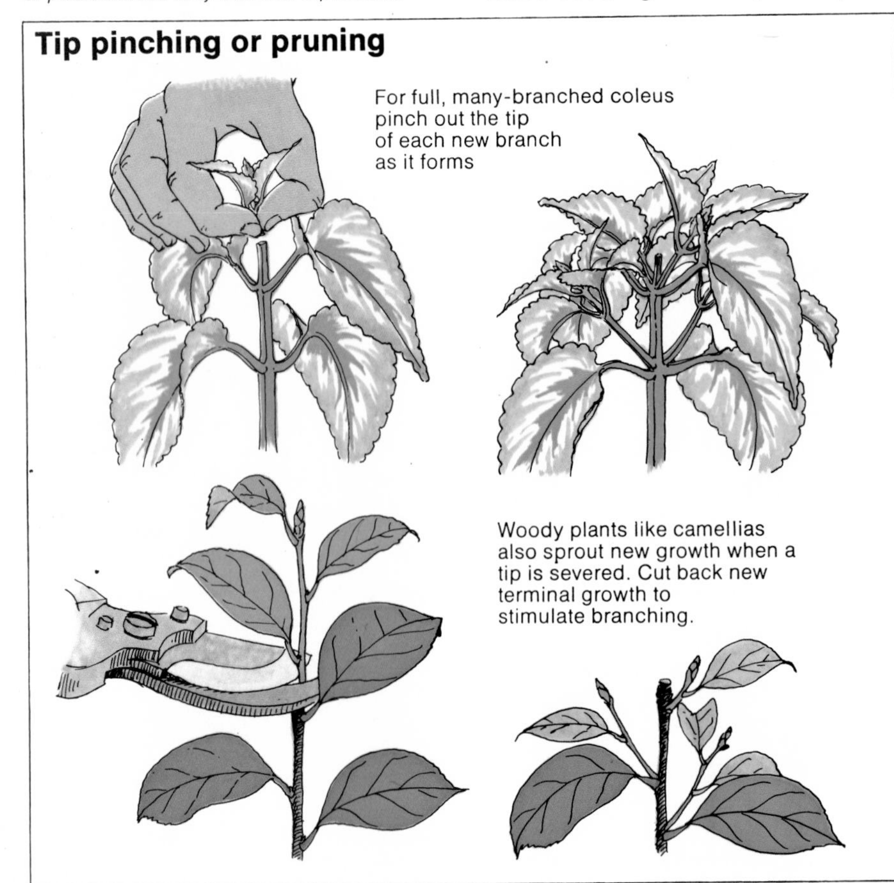

Pest Problems. Insects are not as serious on indoor plants as those grown outdoors. However, they can be brought indoors on new plants. Regular inspection of your plants, as you water, will enable you to spot many pests early enough to save the plant. The following pests are sap feeders and represent those most commonly encountered on house plants.

APHIDS. Clusters of tiny soft-bodied insects usually found in new growth. Foliage is malformed and discolored.

MITES. Look for these tiny 8-legged pests with a magnifying glass. Fine webs may also be seen on undersides of leaves. The plant foliage shows many yellowish specks caused by the sucking of plant juices.

MEALYBUGS. Soft powdery-covered insects look like specks of cotton. Usually found sheltered in axils of stems and on undersides of leaves. Severely affected plants wilt easily and appear yellow and unthrifty.

SCALES. These insects appear as yellowish or greenish brown spots usually on the stems and undersides of leaves. Affected plants turn yellow and show premature leaf drop.

WHITE FLIES. Adults are tiny white winged insects which flutter off the undersides of leaves when plant is disturbed. The most damaging stage of this insect is an almost invisible transparent green larvae that sucks plant juices. Green leaves turn yellow and drop.

If an infested plant is found, it should be isolated from the others and treated as soon as possible with an appropriate indoor houseplant insecticide. Read the label carefully for instructions.

Regular washing of both sides of a shiny foliage plant's leaves will help prevent problems before they occur. Do not wash hairy leaved plants. Any new plant brought into the home should be dipped in a mild soapy water solution or wiped with a small paint brush dipped in a mild detergent solution and isolated and observed for a few days to be sure it's disease free before being added to the plant collection.

Propagation. The indoor garden should include an area for starting new plants from seeds or cuttings. Many soft-stemmed house plants root successfully in water and can be transplanted into a soilless mix shortly after roots have developed or root a

Common house plant problems

Many common house plant problems occur with faulty culture; however, similar symptoms can result from varied causes

Possible causes	Foliage: tips or margins brown	Foliage: bends down and curls	Foliage: yellowish	Foliage: oldest drop	Foliage: all drops	Foliage: spots	Foliage: wilt	Growth: long and weak, thin and soft	Growth: new leaves small	Growth: none develops	Growth: stunted	Growth: plant died
EXCESS LIGHT, i.e. exposure to direct sun can be too intense for many plants.	●	●	●			●		●				
INSUFFICIENT LIGHT impairs photosynthesis.			●└┘	●└┘	●			●		●**		
HIGH TEMPERATURE especially at night, reduces growth and vigor.	●		●	●					●			
LOW TEMPERATURE especially continued exposure, is adverse to the growth of plants.	●			●						●		●
OVERWATERING OR POOR DRAINAGE reduces soil aeration: roots die or water and nutrients are not absorbed.	●		●└┘	●└┘	●		●	●	●		●	●
LACK OF WATER is a most limiting factor of growth.	●		●└┘	●└┘	●	●	●		●*	●		●
LOW HUMIDITY adversely affects the overall health and vigor of most plants.	●				●				●			
TOO MUCH FERTILIZER becomes toxic and injures plant roots.	●		●└┘	●└┘			●		●			●
LACK OF FERTILIZER causes a deficiency of nutrients required for plant growth.	●		●					●	●*			
COMPACTED SOIL reduces root growth and activity.			●				●	●	●		●	
ROOT DECAY from soil-borne insect or disease pests.			●└┘	●└┘	●						●	
DRAFTS cause rapid water loss from foliage.	●		●└┘	●└┘	●							
AIR POLLUTION including an excess of manufactured or natural gas from faulty appliances.		●	●└┘	●└┘					●			

└┘ Cases when low leaves first turn yellow and fall, injury usually progresses up the stem.
*With short internodes on the stem. **With long stem internodes on the stem

cutting directly in any growing medium (soilless mix, vermiculite, sphagnum moss) that is kept moist. Cover with plastic for a few days to keep humid. Repot into soil mix after roots fill the container.

Learn to multiply your plants by sharing cuttings, seeds, or divisions with other gardeners. Not only will you save money but you'll be able to add interesting species to your collection.

Temperature. Most plants will grow in a temperature range of 65° to 75°. A few degrees above or below shouldn't be harmful. Most house plants do best with 50° to 60° night temperatures, therefore it's best to turn the thermostat down at night. However, a few tropical plants, from low elevation habitats (to 1500 ft.) such as Vanda orchids, Fittonias, Guzmanias, etc., may suffer when temperatures go much below 60°. Plants such as Camellia, Cyclamen and Freesia like it cool. For these you may be able to keep a sunny bedroom cool. Or, perhaps you have a sun porch that is cool—but doesn't freeze—in winter. See temperature requirements in charts on pages 28-41.

When light strikes an object, part of its energy is dissipated as heat. Up to a point, plants need heat as well as light. However most problems result from too much heat. Remember, the warmer the temperature, the more humidity plants need to compensate for loss of moisture through evaporation. Plants trapped in a closed southwestern bay window or an unvented skylight may fry on a sunny afternoon unless some sort of shading is provided.

Where winters are severe, tender plants close to windows may freeze. At night, put a piece of cardboard between the plants and the window, or move them to a warmer part of the room. Also, be sure to keep plants out of direct drafts of hot or cold air.

For more growing specifics, please refer to Ortho's HOUSE PLANTS INDOORS/OUTDOORS.

Plants and light

The nature of light and how it affects plants. Photosynthesis, phototropism and photoperiodism. Measuring light.

House plants come to us from all over the tropical and subtropical world. Most of the plants that we grow in our homes are from the cool tropics. Those from the warm tropics require greenhouse culture. Within the tropical belt there is a tremendous variety of lighting conditions, from the dimly-lit understory of dense, tropical jungles to the savannas (grasslands with a few trees and *seasonal* rains) that are open to strong sunlight. Plants differ in their light-intensity and light-duration needs. When a plant is transported from its native habitat to an indoor environment, these differences in lighting requirements must be respected or the plant will not thrive.

How light affects plants

Let's talk a little about light. Light is an essential part of the recipe for healthy plant life. A general understanding of its nature as it relates to indoor gardening will help us keep the cymbidium orchid in bloom and the fiddle-leaf fig from getting sunburned leaves.

Light is radiant energy. Getting technical for a moment, it is the visible portion of the electromagnetic wavelength spectrum. This spectrum is composed of rays of varying wavelength and frequency, and in addition to visible light, there are invisible solar radiations such as X-rays, radio and TV waves and short waves.

Visible 'white' sunlight is really a blend of red, orange, yellow, green, blue and violet rays — all the rainbow hues seen in a schoolboy's glass prism. In the spectrum beyond the visible blue-violet rays are invisible ultra violet rays; at the other end of the spectrum, invisible infrared rays lie just beyond the visible red rays. The various colored rays which make up white light have differing effects on plants. The blue and violet rays promote foliage growth. Plants grown with blue light alone tend to be compact and with lush, dark green leaves, but with few flowers. Research has not revealed any major effects of yellow or green rays on plant growth.

A photoreaction of red and far-red light affects several growth processes in plants. Among them are the elongation and expansion of various plant parts and, notably, flowering. Although plant scientists have been able to identify the effects of this photoreaction, they've yet to discover how they occur.

The research of the United States Department of Agriculture Experimental Station in Beltsville, Maryland, is unearthing the true relationship of light quality to plant life. Application of this knowledge has been made in the development of special 'growth

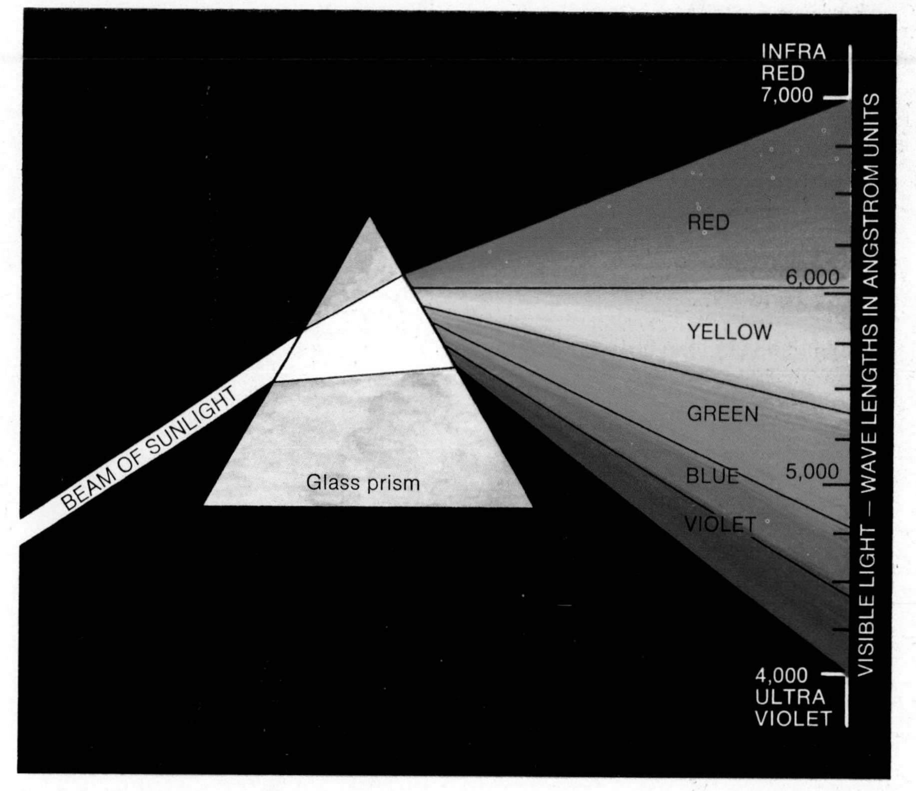

A prism allows us to see the color waves that comprise light; those on the far ends being most important to plants.

◁ *Light provides the radiant energy necessary for plants to manufacture food for growth.*

Growth chambers with controlled lighting have been used successfully in studies at the USDA Experimental Station, Beltsville, Maryland.

The research headed by Dr. H. M. Cathey at the USDA Experimental Station is shedding more light on the effects of various types and intensities of plant light.

lamps,' as we shall see in the chapter covering artificial lighting (page 82).

Some plants are often given too much light—particularly the ones from tropical rain forests. Many of these plants can't even tolerate filtered sunlight—they grow better in a northern window than in a semi-shaded eastern one. If placed in the sun their leaves may wilt during the hottest part of the day, curl downward and develop brown, burned spots. The foliage may undergo color change. Lush greens bleaching out to unhealthy yellows, which happens with scheffleras, philodendrons, ferns and peperomias. Beyond color change is outright leafburn. The 'spider plant' (Chlorophytum), among others, will burn in too much direct sun.

Any plant, regardless of its potential light tolerance, should never be subjected to drastic change of light without conditioning. A plant having been grown in shade can suffer fatal burning if brought directly into a sunny situation. The results can be compared to a person, pale from winter indoor living, going to the beach and getting severely burned while spending many hours in the sun. Unless you know that a plant has already been growing in bright light, it should be gradually exposed to more light, over a period of three to four weeks. Introducing more light daily for gradually longer periods, being watchful of any signs of damage. While some effects of too much sun can be offset by increased watering and humidity, the obvious solution is to move the plant to a more subdued light.

On the other hand insufficient light will produce long, weak stems and leaves and less foliage than normal as the plant stretches toward the light. Scientists call this extended stem growth *etiolation.* Low light intensity will inhibit plants from blooming. A plant surviving at the minimum illumination level only *maintains* itself; if it receives more light, it *grows.* Below the minimum illumination it *weakens* and may die. A plant in this state may appear healthy for some months, but in fact it is utilizing stored carbohydrates and is slowly declining. It may not regain its vigor even if it is again given adequate light.

Most flowering plants *require* direct light. They should be positioned so that they receive direct sunlight part of each day or supplemental artificial light if direct light is not possible. In the winter they can tolerate south and west windowsill treatment in most parts of the United States, and in summer can invariably take direct sun in early morning east windows or late afternoon west ones.

Some plants produce more flowers if the light intensity or the duration is increased. African violets are a notable example; they will produce up to twice as many blooms if they have 16 hours of light rather than 8 hours, assuming a constant 800 foot candles. Most plants do well in more light than their optimum, but few can tolerate less light without serious consequences.

Light and growth inter-actions

Light strongly regulates three major plant processes: Photosynthesis, phototropism and photoperiodism.

PHOTOSYNTHESIS. Light is life. It is one of the necessary ingredients for photosynthesis (from *photo* = light and *synthesis* = to put together). It provides the energy required for the manufacture of a plant's food. Water and atmospheric carbon dioxide are converted in the presence of light into carbohydrates. With these carbohydrates the plant produces new growth: foliage, roots, stems and blooms. The presence of optimum light fosters luxuriant leaf growth and flowering based on the right light for the right plant.

Photosynthesis stops with the absence of light. In the artificial atmosphere indoors a plant's growing 'season' can be greatly extended by artificial light.

PHOTOTROPISM is the natural tendency of plants to grow toward their light source.

Phototropism (from *photo* = light and *trope* = to bend) is controlled by an auxin, or growth hormone, which occurs at the stem tips and youngest plant leaves. This auxin is highly reactive to light and causes the plant to unceasingly adjust itself to the light source. Positive tropism is a condition where a plant moves *toward* the light source (the most common situation) and negative tropism occurs where a plant moves *away* from the source.

Indoors where the natural light source is normally a window, plants will bend toward the window. In order to avoid developing permanently-misshapen plants rotate them occasionally to foster uniformly upright growth. Plants that grow rapidly need to be turned more often.

Artificial lighting should come from above—at least shining on the top half of an erect plant. The best source is from directly overhead. Weeping plants such as a hanging Boston fern, can be successfully lighted from below, but the light must be of much greater intensity than when the light source is above the plant.

PHOTOPERIODISM. All plants are light-programmed to their own native environment and perform best in the rhythmic light/darkness cycle found there. For many plants the length of nights and days is a determining factor in the time required to reach maturity—that stage in a plant's life when reproduction becomes possible.

Some plants flower best when the days are long—14 hours or more; these are called *long-day* plants. Some long-day plants are calceolaria, tuberous begonia, cineraria, hibiscus, heather and nasturtium.

Conversely, other plants produce blooms when days are short; these are called *short-day* plants. Short-day plants will not form flower buds unless they receive at least 14 hours of darkness while their flower buds are setting, prior to bloom. Common short-day plants are gardenias, kalanchoe, chrysanthemum, Christmas cactus, poinsettia, cattleya orchids and bougainvillea, among others.

Most plants however have no definite response—they are *day-neutral.* They will bloom, generally on schedule, in either 8 or 16 daily hours of light. Bulbs are day-neutral; tulips, daffodils and amaryllis all seem to perform well with differing day lengths.

Armed with the knowledge that fourteen or more hours a day of light triggers flowering in long-day plants, indoor gardeners can force blooming any time of the year by using artificial light. On the other hand, many short-day plants requiring twelve hours or more of darkness to stimulate flowering can be induced into bloom artificially by putting them in a light-tight closet at night and taking them out the next day at noon, or by covering them with a photographer's black cloth or other light-shielding material.

The seasonal dormant cycle of outdoor plants is missing for house plants because the home environment does not have the periodic shifts in temperature and other natural impacts that a plant is accustomed to in its natural habitat. A plant will be healthiest if we can duplicate its natural environment. Dormancy in woody plants can be induced by short days and reduced watering. Although it seems like a lot of work to decrease light, water and feeding during a plant's 'winter,' we know that most plants require a rest for healthy spring growth and, indeed, some flowering plants such as gloxinia, ixia and tuberous begonia need a period of dormancy if they are to bloom at all.

This Syngonium *is a victim of phototropism as it is forced to reach out for more light.*

Measuring light

Light is measured in footcandles and lumens, depending on whether one is considering the object that is lighted or the source of the light.

Footcandles, or f.c. for short, are the amount of light *received* on a surface, and *lumens* are the measurement of light *emitted* by a light source.

Natural sunlight and artificial light falling on a plant are measured in footcandles while the light emitted by such sources as sun itself and electric lamps are rated in lumens.

One footcandle is the amount of visible light falling on one square foot of surface located one foot away from one candle. One lumen is the amount of light given off by one candle that falls on one square foot of surface one foot away.

For an example of common measurements:

✓A footcandle meter reading at noon on a clear summer day may reach 10,000 f.c.

✓A mid-day reading on an overcast winter day may be as low as 500 f.c.

Inside, light readings are much lower. The light may register 4,000-8,000 footcandles in the direct sun entering a window on a clear summer afternoon, while directly in the shade to the side of the window the meter may indicate 600. At the same time of day on the shady side of the house the window reading might be expected to be 150-250 f.c.

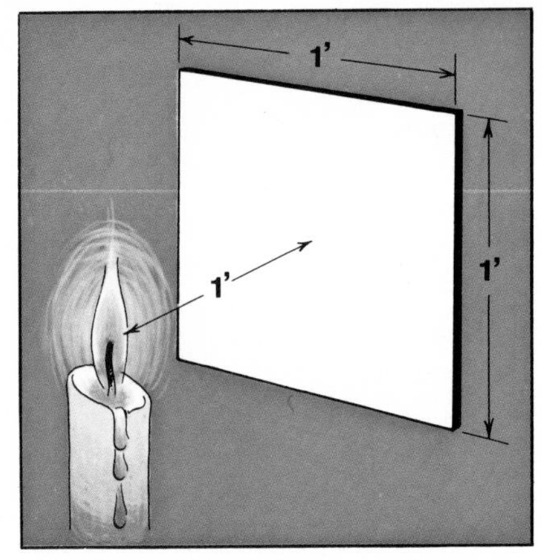

A footcandle *is the amount of light falling on one square foot of surface located one foot away from one candle.*

Lamps are rated in lumens. A 40-watt rapid-start fluorescent tube generally casts in the range of 2,000-3,000 lumens initially, normally decreasing with the life of the lamp.

It is difficult to tell true light intensity 'by eye' because our eyes automatically adjust themselves to everything that they see! A bright summer day with light fog casting no discernable shadows may register higher inside on the footcandle scale than a clear winter afternoon when there is strong light-shade contrast and an indoor plant is not in a direct-sun situation. This is true because of the reflective quality of gray or overcast skies as opposed to the light absorption of blue skies. On the other hand, the back of a dark room will probably register lower on a light meter than would seem likely, because our eyes have adjusted themselves to the relative darkness.

Some indoor gardeners are so experienced at their hobby that they have a built-in sense of available light. They will instinctively know if a certain spot has too much direct sun for a false aralia, or not enough for a flowering maple. If only we all possessed this built-in light meter, we could dispense with the necessary step of estimating available light. However, most of us don't have this innate ability, so we must resort to other methods.

The most accurate way to estimate light is to employ mechanical means. This can be done with a special light meter which reads directly in footcandles, or we can utilize a photographer's light meter or the light meter that is built into a camera and translate the photographic readings.

A light meter with a scale in direct footcandle readings is manufactured by the General Electric Company. It is Model #214, and records up to 1,000 f.c.

In order to make a reading with the footcandle meter, place the meter at the same position as the surface of the leaves. Aim the plastic-covered lens toward the maximum light source. Without blocking the light or casting a shadow on the meter, check the reading on the dial. The reading will be accurate to within 10-15%, which is adequate for our purposes.

A photographic meter or camera with a built-in light meter will pro-

This light meter registers the number of footcandles falling on the piggyback (Tolmeia menziesii). *It's the easiest method of measuring actual light on your plants.*

vide fairly accurate readings that can be translated into footcandles. Here are two methods:

✔Method 1 (illustrated below)
Set the film-speed dial to ASA 100 and aim the camera or hand held meter at a sheet of matte white cardboard or paper in the proposed plant location and orient to the maximum light source. Get close enough to the paper so the meter sees only the white paper. Be sure not to block the light or create a shadow. The shutter speed indicated opposite stop f4, reading it as a whole number, will be the approximate footcandles of illumination measured. For example, if the f-stop registers an exposure of 1/250 second, there are about 250 f.c. of light playing on the white sheet.

✔Method 2 (not illustrated)
Set the ASA film-speed at 200 and the shutter speed at 1/125 second. Focus on the white paper as above. Adjust the f-stop until a correct exposure is shown in the light meter in camera. By using the table below, the f-stop (lens opening) will tell you how many footcandles you have.

f2.8	32 f.c.
f4	64 f.c.
f5.6	125 f.c.
f8	250 f.c.
f11	500 f.c.
f16	1000 f.c.
f22	2000 f.c.

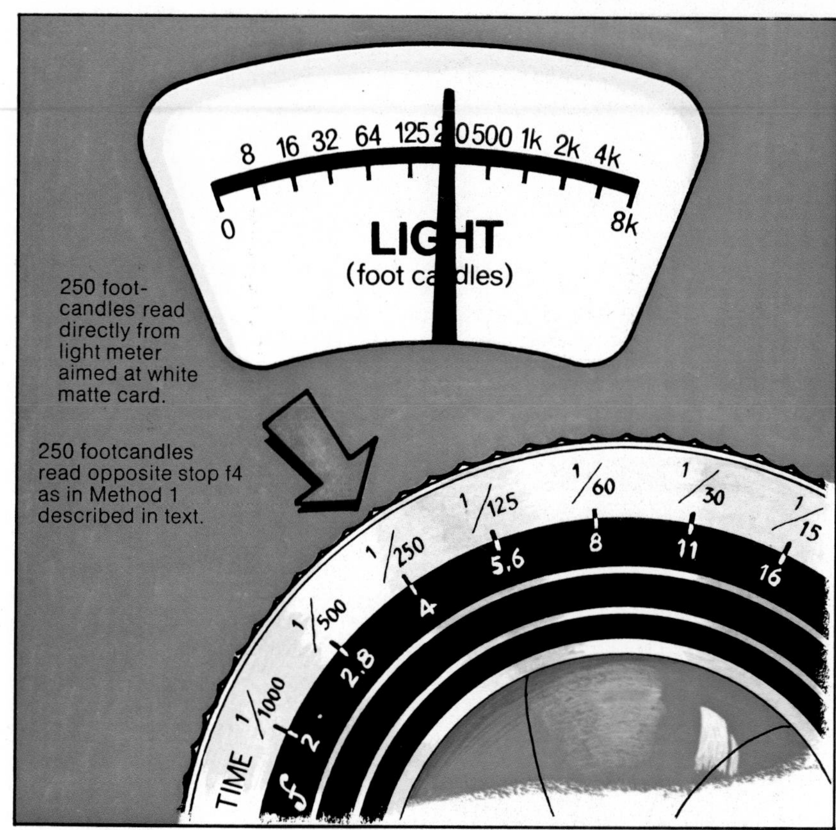

Photographic light meters or built-in camera meters can be used to measure footcandles by the two methods described above.

The camera's photographic meter measures the same spectrum of visible light as the footcandle meter. The latter registers intermediate readings and is more accurate as well.

A third method for estimating light is a non-mechanical one and we have done most of the work for you—just follow the diagrams for each exposure on pages 46-52. This technique is not nearly as accurate as direct measurement, but it will be sufficient for the average homeowner or apartment dweller, who is primarily interested in foliage plants and may not have the available equipment to employ the light meter approach.

LOW INTENSITY. Even the most shade tolerant of houseplants barely exist in a dark northern corner with little light. This should be considered the absolute minimum intensity for even the 'maintenance' level of existence. Snake plants (sansevieria), philodendron, dieffenbachia, dracaena, syngonium, chamaedorea and pothos will survive on extremely low light intensities (50-100 f.c.). Variegated foliage frequently loses all or part of its variegation at these minimum levels.

Many rooms receive too little natural light for any plants to survive. Don't be fooled into thinking that the plants you see thriving in hotel lobbies, offices, restaurants and department stores are receiving the light they need for growth. Behind the scene such plants are given special treatment. A plant rental service, or a faithful employee gives the plants the ideal light-water-temperature-nutrient treatment before placing them in the unfavorable environment. They apparently thrive there but they are living on their reserves and will be replaced by fresh plantings before they show signs of stress. Rotation of plantings on a regular schedule is the secret.

Generally speaking the less light a plant gets, the lower the temperature should be. For this reason poorly-lit areas should be kept as cool as possible. Plants in these situations also require less water and should be fertilized less often than their brightly-lit counterparts.

The length of darkness has little effect on foliage plants—whether a plant gets twelve hours of darkness or only six hours seems to make little difference. In fact if your heart-leaved Philodendron isn't getting enough light, leave the lights on longer. A shorter dark period will do it no harm and longer illumination will at least partially offset lack of intensity. A nearby lamp turned on each evening may very well be enough to maintain vigorous growth as long as it does not produce excessive heat.

If artificial light is the only source of light and a foliage plant is receiving only the 'maintenance' level of illumination—say in a 12 hour day, then the lamps should be on a 'long-day' every day—16 hours.

In many cases artificial light is used as a supplement to natural light. In these instances it is best to have the lights on in the *daytime* to increase the intensity into the proper bracket and provide a more natural cycle, leaving some dark hours.

HIGH INTENSITY LIGHT. Flowering plants require much higher light intensities than foliage plants. Some perform on a minimum of 800 f.c. through twelve hours, while most, including the majority of annuals and perennials require at least 1,000 f.c.

Seedlings and cuttings need high intensity light. They can be started in late winter under 40 watt fluorescent tubes and be ready to be set in the ground when spring arrives. Place the tubes as close as two inches above the sprouting seedlings and raise on chains as they grow up. Fast-growing plants such as tomatoes, zinnias, marigolds and cucumbers need a minimum of 16 hours a day of 1,000-1,500 f.c. for growth to transplant size.

Plant selection guide

These charts should be used only as a guideline. Higher humidity with good air circulation, for example, can permit admission of more intense light than the charts might indicate; but extremely low humidity with still air could sharply drop the tolerance level. Always make changes progressively, moving one step at a time from low light intensity to bright, direct light. If a sudden adverse change results, take one step back and allow a longer period before trying the next rung on the ladder to higher light intensity.

Light requirements: *Direct sun* — sunlight falls directly on the plant for as long as 4-6 hours per day; above 800 f.c. *Winter direct sun* — sunlight falls directly on the plant in winter only. *Bright* — no direct sunlight on the plant, but intense indirect or filtered sunlight; 400-800 f.c. *Moderate* — an intermediate amount of light; 250-400 f.c. *Low* — a low amount of light, generally away from windows; 100-250 f.c. *Very low* — the lowest amount of light any plant can stand and still live; 50-100 f.c.

Exposures: *North* — a "window" facing north or more north than any other direction. *East* — a "window" facing east or more east than any other direction. *West* — a "window" facing west or more west than any other direction. *South* — a "window" facing south or more south than any other direction.

Water requirements: *Keep wet at all times* — never allow the plant to dry out. Keep standing in saucer of wet gravel, changing every few days to avoid stagnation. *Keep evenly moist* — growing medium remains nicely moist to the touch, never oozing with water or dusty dry. *Approach dryness between waterings* — soaking the rootball, then allowing soil to begin to feel dry to the touch before repeating soaking. *Dry out between waterings* — periods of soaking between which the soil should be allowed to become completely dry.

Humidity: *Average house* — 30-45 percent. *Moist* — 45-60 percent. *Very moist* — above 60 percent.

Temperature: *Cool*—Day—55-60° F. (13-15° C.). Night—40-50° F. (5-7° C.). *Average house* — Day — 70° F. (21° C.) or slightly above. Night — 50-55° F. (10-13° C.). *Warm* — Day — 80-85° F. (27-30° C.). Night — 62-65° F. (16-18° C.).

COMMON NAME	BOTANICAL NAME	LIGHT REQUIREMENTS						EXPOSURE				WATER REQUIREMENTS				HUMIDITY			TEMPERATURE			MOST ATTRACTIVE FEATURES		
		Direct sun	Winter direct sun	Bright	Moderate	Low	Very low	North	East	West	South	Keep wet at all times	Keep evenly moist	Approach dryness between waterings	Dry out between waterings	Average house	Moist	Very moist	Cool	Average house	Warm	Foliage	Flowers	Fruit
Achimenes, Rainbow flower	*Achimenes* species	•		•					•	•	•		•				•			•			•	
African hemp	*Sparmannia africana*			•					•	•	•		•				•			•		•		
African violets	*Saintpaulia ionantha*		•	•	•				•	•			•				•			•		•	•	
Agapanthus	*Agapanthus* species	•		•	•				•	•	•		•			•			•	•		•	•	
Airplane plant (see Spiderplant)																								
Allamanda	*Allamanda* species	•							•	•	•		•				•			•		•	•	
Aluminum plant	*Pilea cadierei*			•	•				•	•			•				•			•		•		
Amaryllis	*Hippeastrum vittatum*	•		•					•	•	•		•				•			•		•	•	
Amazon lily	*Eucharis grandiflora*		•	•					•	•	•		•				•			•			•	
Annuals and Vegetables	Many species	•							•	•	•		•			•				•	•	•	•	•
Anthurium, Flamingo flower	*Anthurium* species			•	•				•				•				•		•			•	•	
Aphelandra (see Zebra plant)																								
Apostle plant	*Neomarica gracilis*		•	•					•	•			•			•			•			•	•	
Arabian violet, German violet	*Exacum affine*		•	•					•				•				•			•			•	
Aralia, Fatsia (also see *Dizygotheca* and *Polyscias*)	*Fatsia japonica*		•	•	•				•	•	•		•				•			•		•		
Ardisia, Coral berry	*Ardisia* species			•					•	•			•				•		•			•		•
Artillery plant	*Pilea microphylla*				•				•	•			•				•		•			•		
Asparagus fern	*Asparagus* species			•	•	•		•	•	•	•		•				•			•		•		
Aspidistra	*Aspidistra elatior*				•	•	•	•	•	•			•			•			•			•		
Aucuba	*Aucuba japonica*			•	•				•	•			•			•			•			•		
Avocado	*Persea americana*	•		•					•	•	•		•				•			•		•		
Azalea	*Rhododendron* species			•					•		•		•					•	•			•	•	
Aztec lily, Jacobean lily	*Sprekelia formosissima*	•								•	•			•		•					•		•	
Baby's tears	*Helxine soleirolii*				•	•		•	•				•				•	•		•		•		
Bamboo	*Bambusa* species	•								•	•	•					•			•		•		
Banana	*Musa* species		•						•	•	•	•					•				•	•		
Barbados cherry	*Malpighia glabra*	•							•	•				•		•				•			•	
Basket vine	*Aeschynanthus speciosus*			•					•	•			•			•				•			•	
Begonia: Fiberous-rooted	*Begonia* species			•					•	•	•		•				•			•		•	•	
Rex	*Begonia* Rex-cultorum			•	•				•	•			•				•		•			•	•	
Wax	*Begonia semperflorens*	•							•	•	•		•				•		•				•	

Blooms of Lipstick vine (Aeschynanthus *species*) *are contrasted with delicate fronds of* Nephrolepis exaltata 'Elegantissima'.

COMMON NAME	BOTANICAL NAME	LIGHT REQUIREMENTS						EXPOSURE				WATER REQUIRE-MENTS				HUMIDITY			TEMPERA-TURE			MOST ATTRAC-TIVE FEATURES		
		Direct sun	Winter direct sun	Bright	Moderate	Low	Very low	North	East	West	South	Keep wet at all times	Keep evenly moist	Approach dryness between waterings	Dry out between waterings	Average house	Moist	Very moist	Cool	Average house	Warm	Foliage	Flowers	Fruit
Bird of paradise	*Strelitzia* species	●		●					●		●		●			●			●			●	●	
Black-eyed Susan vine	*Thunbergia alata*	●							●		●		●			●				●			●	
Bleeding heart (see *Clerodendrum*)																								
Blood lily	*Haemanthus* species	●		●					●		●		●				●			●			●	
Blue Marguerite (see *Felicia*)																								
Bougainvillea	*Bougainvillea glabra*	●								●	●		●				●			●			●	
Brazilian edelweiss	*Sinningia leucotricha*		●							●				●			●				●	●		
Bromeliads: Living vase plant	*Aechmea* species			●					●	●				●			●			●		●	●	
Pineapple	*Ananas* species	●								●	●			●			●	●		●		●		●
Queen's Tears	*Billbergia* species		●	●					●	●			●				●			●		●	●	
Volcano plant	*Bromelia* species	●								●			●				●			●		●	●	●
No common name	*Catopsis* species	●									●		●				●			●		●	●	
Earth stars	*Cryptanthus* species			●	●				●	●			●				●	●		●		●		
No common name	*Dyckia* species			●	●				●	●				●		●				●		●		
No common name	*Guzmania* species		●						●		●		●				●			●		●	●	
Fingernail plant	*Neoregelia* species		●	●					●	●				●			●	●		●		●		
No common name	*Nidularium* species			●					●	●			●				●				●	●		
No common name	*Portea* species	●		●	●				●	●				●			●				●	●	●	
Grecian vase	*Quesnelia* species			●					●	●			●			●				●		●	●	
No common name	*Tillandsia* species	●		●					●	●					●		●					●	●	
Flaming sword	*Vriesea* species				●	●			●	●			●	●			●			●		●	●	
Butterfly gardenia	*Tabernaemontana divericata*	●								●	●			●		●				●			●	
Cacti: Rat-tail	*Aporocactus* species	●		●						●	●					●				●		●	●	
Bishop's cap	*Astrophytum* species	●		●						●	●					●				●		●	●	
Old man	*Cephalocereus* species	●		●						●	●					●				●		●		
No common name	*Cereus* species	●		●						●	●					●				●		●	●	●
Column/Torch	*Cleistocactus* species	●		●						●	●					●				●		●	●	
Barrel/Star	*Echinocactus* species	●		●						●	●					●				●		●	●	
Rainbow/Hedgehog	*Echinocereus* species	●		●						●	●				●	●				●		●	●	●
Barrel	*Echinopsis* species	●		●						●	●				●	●				●		●	●	●
Orchid cactus	*Ephiphyllum* species			●	●				●	●	●		●				●			●		●	●	
Chin	*Gymnocalycium* species	●		●						●	●					●				●		●	●	
Night-blooming	*Hylocereus* species			●	●				●	●	●		●				●				●	●	●	
Cob	*Lobivia* species	●		●						●	●					●				●		●	●	
Pincushion	*Mammillaria* species	●		●						●	●					●				●		●	●	
Ball	*Notocactus* species	●		●						●	●					●				●		●	●	
Bunny ears	*Opuntia* species	●		●						●	●					●				●		●	●	●
Lemon vine	*Pereskia* species	●		●						●	●					●				●		●	●	
Crown	*Rebutia* species	●		●						●	●					●				●		●	●	
Mistletoe cactus	*Rhipsalis* species				●	●		●	●				●				●			●		●	●	●
Christmas cactus	*Schlumbergera* species			●	●				●	●	●		●				●			●		●	●	
Thanksgiving cactus	*Zygocactus* species			●	●				●	●	●		●				●			●		●	●	
Caladium	*Caladium* species			●	●			●	●	●			●				●				●	●		
Calamondin (see Citrus)																								
Calceolaria	*Calceolaria* species				●	●		●	●				●				●		●				●	
Calico flower	*Aristolochia* species			●					●	●			●			●				●			●	
Calla lily	*Zantedeschia* species	●							●		●	●					●			●			●	
Camellia	*Camellia* species			●						●	●		●					●	●			●	●	
Campanula, Star-of-Bethlehem	*Campanula* species			●					●	●	●		●			●			●				●	

A collection of Cacti combine effectively with Dracaena warnecki. ⇨

COMMON NAME	BOTANICAL NAME	LIGHT REQUIREMENTS						EXPOSURE				WATER REQUIRE-MENTS				HUMIDITY			TEMPERA-TURE			MOST ATTRAC-TIVE FEATURES		
		Direct sun	Winter direct sun	Bright	Moderate	Low	Very low	North	East	West	South	Keep wet at all times	Keep evenly moist	Approach dryness between waterings	Dry out between waterings	Average house	Moist	Very moist	Cool	Average house	Warm	Foliage	Flowers	Fruit
Candle plant	*Plectranthus oertendahlii*				•				•	•			•				•			•		•	•	
Cape primrose	*Streptocarpus* species	•		•					•	•	•			•		•				•			•	
Cardinal flower	*Sinningia cardinalis*			•						•	•			•		•				•			•	
Carolina jasmine	*Gelsemium sempervirens*			•					•	•			•				•			•			•	
Cast iron plant (see *Aspidistra*)																								
Century plant (see Succulents)																								
Chenille plant	*Acalypha hispida*	•							•	•	•		•			•				•			•	
Chincherinchee	*Ornithogalum thyrsoides*	•		•					•		•			•		•			•				•	
Chinese evergreen	*Aglaonema* species				•	•	•	•	•			•					•			•		•		
Christmas cherry, Jerusalem cherry	*Solanum pseudocapsicum*	•		•					•				•				•		•					•
Chrysanthemum	*Chrysanthemum* species			•					•				•			•			•				•	
Cigar plant	*Cuphea ignea*	•		•					•	•	•		•				•			•		•	•	
Cineraria	*Senecio hybridus*			•	•				•	•			•				•		•				•	
Citrus:																								
Chinese lemon	*Citrus limon 'Meyer'*	•							•	•	•		•			•				•		•		•
Wonder lemon	*Citrus limon 'Ponderosa'*	•							•	•	•		•			•				•		•		•
Calamondin	*Citrus mitis*	•								•	•		•			•				•		•		•
Kumquat	*Fortunella margarita*	•								•	•		•			•				•		•		•
Tangerine, Satsuma orange	*Citrus reticulata*	•								•	•		•			•				•		•		•
Sweet orange	*Citrus sinensis*	•								•	•		•			•				•		•		•
Otaheite orange	*Citrus taitensis*	•								•	•		•			•				•		•		•
Clerodendrum, Glory-bower, Bleeding heart	*Clerodendrum* species			•					•	•	•		•			•	•			•		•	•	
Clivia	*Clivia* species			•					•		•		•			•			•	•			•	
Cobra lily	*Darlingtonia californica*				•	•	•	•	•			•					•			•		•		
Coffee	*Coffea* species			•					•		•	•					•		•	•		•	•	
Coleus	*Coleus* species	•		•	•				•	•	•		•			•	•			•		•		
Columnea	*Columnea* species			•					•	•			•				•	•		•			•	
Copperleaf	*Acalypha wilkesiana*	•		•					•		•		•			•				•		•		
Coral plant	*Russelia equisetiformis*	•		•					•					•			•			•			•	
Corn plant (see *Dracaena*)																								
Creeping-Charlie	*Pilea nummulariifolia*				•			•	•	•			•				•			•		•		
Creeping fig (see Figs)																								
Crinum	*Crinum* species		•	•					•	•	•		•				•			•		•	•	
Crocus	*Crocus* species	•		•					•		•		•			•			•				•	
Crossandra, Firecracker flower	*Crossandra infundibuliformis*		•						•	•			•				•			•		•	•	
Croton	*Codiaeum* species	•							•				•				•			•		•		
Cup-and-saucer vine	*Cobaea scandens*			•					•		•		•				•			•		•	•	
Cycas	*Cycas* species	•		•				•	•	•			•				•			•		•		
Cyclamen	*Cyclamen persicum*			•					•				•					•	•			•	•	
Cyperus	*Cyperus* species		•	•					•	•	•	•					•		•			•		
Daffodils	*Narcissus* varieties			•					•	•	•		•			•			•				•	
Devil's backbone	*Pedilanthus tithymaloides*			•	•			•	•	•			•				•			•		•		
Devil's ivy (see Pothos)																								
Dieffenbachia	*Dieffenbachia* species			•	•	•		•	•					•			•			•		•		
Dipladenia	*Dipladenia* species	•		•					•	•	•				•		•	•		•			•	
Dizygotheca	*Dizygotheca elegantissima*			•	•			•	•				•				•			•		•		
Double decker plant	*Sinningia verticillata*		•						•	•				•			•				•		•	
Dracaena	*Dracaena* species				•	•	•	•	•	•		•					•			•		•		

COMMON NAME	BOTANICAL NAME	LIGHT REQUIREMENTS						EXPOSURE				WATER REQUIREMENTS				HUMIDITY			TEMPERATURE			MOST ATTRACTIVE FEATURES		
		Direct sun	Winter direct sun	Bright	Moderate	Low	Very low	North	East	West	South	Keep wet at all times	Keep evenly moist	Approach dryness between waterings	Dry out between waterings	Average house	Moist	Very moist	Cool	Average house	Warm	Foliage	Flowers	Fruit
Dragon tree (see *Dracaena*)																								
Easter lily	*Lilium longiflorum*			●					●		●			●		●			●			●		
Egyptian star flower	*Pentas lanceolata*	●							●	●	●		●			●				●			●	
Elephant ears	*Colocasia* species			●					●	●	●	●					●				●	●		
Elephant foot tree (see Ponytail)																								
Episcia	*Nautilocalyx*			●	●				●	●	●		●				●				●		●	
Euonymus (evergreen)	*Euonymus japonica*				●				●	●			●				●		●			●		
False aralia (see *Dizygotheca*)																								
False sea onion	*Ornithogalum caudatum*	●		●					●	●	●			●		●			●				●	
Fatshedera, Tree ivy	*Fatshedera lizei*	●		●	●			●	●	●	●		●				●		●	●		●		
Fatsia (see Aralia)																								
Felicia, Blue Marguerite	*Felicia amelioides*	●		●					●	●	●		●				●			●			●	
Ferns:																								
Bear's paw	*Polypodium polycarpon*		●					●	●				●			●				●		●		
Birdsnest	*Asplenium nidus*					●	●	●	●				●				●		●			●		
Boston	*Nephrolepis exaltata bostoniensis*				●	●		●	●				●				●			●		●		
Brake ferns	*Pteris* species		●	●	●	●							●					●	●			●		
Deersfoot	*Davallia canariensis*					●		●	●				●				●	●	●			●		
Fluffy ruffles	*Nephrolepis exaltata* 'Fluffy Ruffles'				●	●		●	●				●				●			●		●		
Hare's foot	*Polypodium aureum*		●	●	●				●		●		●				●		●			●		
Hart's tongue	*Phyllitis scolopendrium*					●		●	●				●				●		●			●		
Hawaiian tree	*Cibotium chamissoi*				●				●				●				●			●		●		
Hollyfern	*Cyrtomium falcatum*		●	●	●	●		●	●				●				●		●			●		
Maidenhair	*Adiantum* species		●	●	●	●		●	●			●					●	●		●		●		
Mother	*Asplenium bulbiferum*				●	●		●	●				●				●		●			●		
Rabbit's foot	*Davallia fejeensis*				●	●		●	●				●				●	●	●			●		
Squirrel's foot, Ballfern	*Davallia mariesii*				●	●		●	●				●				●	●		●		●		
Staghorn	*Platycerium bifurcatum*			●	●				●	●			●				●			●		●		
Treefern	*Cibotium* species				●				●				●				●			●		●		
	Alsophila species			●	●				●	●	●	●					●				●	●		
	Cyathea species				●				●	●	●	●					●				●	●		
Trembling brake	*Pteris tremula*				●	●		●	●				●				●	●	●			●		
Whitman	*Nephrolepis exaltata* 'Whitmanii'				●	●		●	●				●					●		●		●		
Figs *(Ficus)*:																								
Creeping	*Ficus pumila*			●	●				●	●			●				●			●		●		
Fiddle-leaf	*Ficus lyrata*			●	●				●	●			●			●	●			●		●		
Indian laurel	*Ficus retusa*			●	●				●	●			●				●			●		●		
Mistletoe	*Ficus deltoidea*			●	●				●	●			●			●	●			●		●		●
Rubber plant	*Ficus elastica*			●	●	●		●	●	●	●		●			●	●			●		●		
Weeping	*Ficus benjamina*			●	●				●	●	●		●				●			●		●		
Firecracker flower (see *Crossandra*)																								
Firecracker vine	*Manettia cordifolia 'Glabra'*			●					●				●				●			●			●	
Fittonia	*Fittonia* species					●		●	●				●				●	●	●			●		
Flagplant (miniature)	*Acorus gramineus*				●	●		●	●			●						●	●			●		
Flame-of-the-woods	*Ixora* species	●		●					●	●	●			●			●			●			●	
Flame violet (see *Episcia*)																								
Flamingo flower (see *Anthurium*)																								
Flowering maple	*Abutilon* species	●							●				●			●				●		●	●	

COMMON NAME	BOTANICAL NAME	LIGHT REQUIREMENTS						EXPOSURE				WATER REQUIREMENTS				HUMIDITY			TEMPERATURE			MOST ATTRACTIVE FEATURES		
		Direct sun	Winter direct sun	Bright	Moderate	Low	Very low	North	East	West	South	Keep wet at all times	Keep evenly moist	Approach dryness between waterings	Dry out between waterings	Average house	Moist	Very moist	Cool	Average house	Warm	Foliage	Flowers	Fruit
Flowering tobacco (see *Nicotiana*)																								
Freesia	*Freesia* species			•					•	•	•		•				•		•				•	
Fuchsia	*Fuchsia hybrida* varieties			•					•	•			•				•		•			•	•	
Gardenia	*Gardenia jasminoides*		•								•		•				•			•		•	•	
Gasteria (see Succulents)																								
Gazania	*Gazania* species	•		•					•	•	•			•			•			•			•	
Geranium	*Pelargonium* species	•		•					•	•	•			•		•	•		•	•			•	
Geranium (Ivy)	*Pelargonium* peltatum	•		•					•	•	•			•		•	•		•	•			•	
German ivy (see Parlor ivy)																								
Ginger:	*Alpinia* species		•	•	•				•				•			•	•			•		•	•	
No common name	*Amomum* species		•	•	•				•				•			•	•			•		•	•	
Spiral	*Costus* species		•	•	•				•	•			•				•			•		•	•	
Hidden lily	*Curcuma* species		•	•	•				•	•			•				•	•		•		•	•	
Ginger lily	*Hedychium* species	•							•	•		•	•				•			•		•	•	
Peacock plant	*Kaempferia* species		•	•	•				•	•			•					•		•		•	•	
Common ginger root	*Zingiber* species		•	•					•				•				•	•		•		•		
Glory bower (see *Clerodendrum*)																								
Glory lily	*Gloriosa* species	•		•					•	•	•		•				•			•			•	
Gloxinia	*Sinningia* species			•	•				•				•				•			•			•	
Gloxinera	*Gloxinera* varieties			•					•				•				•			•			•	
Goldust plant	*Aucuba japonica 'Variegata'*			•	•	•		•	•					•		•			•			•		
Goldfish plant	*Nematanthus* species		•						•	•			•			•	•			•		•	•	
Grape-hyacinth	*Muscari* species	•		•					•	•	•	•					•		•				•	
Grape ivy	*Cissus rhombifolia*				•	•		•	•	•			•				•			•		•		
Grecian urn plant, Grecian pattern	*Acanthus* species			•					•	•	•		•			•				•		•		
Gynura	*Gynura scandens*			•	•				•	•			•				•			•			•	
Hawaiian ti	*Cordyline terminalis*			•	•	•		•	•	•	•		•				•			•		•		
Heath, Heather	*Erica* species			•					•	•	•						•			•			•	
Heavenly bamboo (see *Nandina*)																								
Heliotrope	*Heliotropium arborescens*	•							•	•	•		•				•			•			•	
Hibiscus, Chinese hibiscus	*Hibiscus rosa-sinensis*	•		•					•		•		•				•			•		•	•	
Holly osmanthus	*Osmanthus* species	•		•					•	•	•		•				•			•		•		
Hyacinth	*Hyacinthus orientalis*	•		•	•				•	•	•		•				•		•				•	
Hydrangea	*Hydrangea macrophylla*	•		•					•		•	•	•			•			•				•	
Hypoestes	*Hypoestes* species	•		•					•		•		•				•			•			•	
Impatiens	*Impatiens* species		•	•					•	•				•				•		•			•	
Indian laurel (see Figs)																								
Indoor oak	*Nicodemia* species	•		•					•		•		•				•				•	•		
Iresine	*Iresine* species	•		•					•		•			•		•			•	•		•		
Ivy (Algerian)	*Hedera canariensis*	•		•					•	•	•		•			•			•			•		
Ivy (English)	*Hedera helix*			•				•	•		•		•	•		•	•		•			•		
Ixora (see Flame-of-the-woods)																								
Jacobean lily (see Aztec lily)																								
Jacobinia (see King's crown)																								
Jasmine	*Cestrum, Jasminum* species			•					•		•		•			•	•		•			•	•	
Jerusalem cherry (see Christmas cherry)																								
Kaffir lily (see *Clivia*)																								
Kangaroo vine, Kangaroo ivy	*Cissus antarctica*			•	•	•		•	•				•				•			•		•		

◁ *Kangaroo vine* (Cissus antarctica) *is illuminated at night by incandescent back lighting attached to the underside of the bookshelf.*

COMMON NAME	BOTANICAL NAME	LIGHT REQUIREMENTS						EXPOSURE				WATER REQUIRE-MENTS				HUMIDITY			TEMPERA-TURE			MOST ATTRAC-TIVE FEATURES		
		Direct sun	Winter direct sun	Bright	Moderate	Low	Very low	North	East	West	South	Keep wet at all times	Keep evenly moist	Approach dryness between waterings	Dry out between waterings	Average house	Moist	Very moist	Cool	Average house	Warm	Foliage	Flowers	Fruit
Kenilworth ivy	*Cymbalaria muralis*			●					●				●				●		●			●		
King's crown	***Justicia carnea***	●	●	●					●		●		●				●			●			●	
Kohleria	*Kohleria* species			●					●	●			●				●				●		●	
Kumquat (see Citrus)																								
Lantana	*Lantana* species	●		●					●	●	●			●		●			●				●	
Lemon (see Citrus)																								
Leopard plant	*Ligularia tussilaginea aureo-maculata*					●	●	●	●				●				●		●			●		
Lily of the Nile (see *Agapanthus)*																								
Lily-of-the-valley	*Convallaria majalis*	●							●	●		●				●			●				●	
Lily turf (see Liriope)																								
Lipstick vine	*Aeschynanthus* species			●					●		●		●				●	●		●			●	
Liriope, Lily turf	***Liriope* species**			●	●				●	●	●		●				●		●				●	
Living stones (see Succulents)																								
Loquat	*Eriobotrya japonica*	●		●					●	●	●		●				●			●		●		
Love plant	*Medinilla magnifica*			●	●				●	●			●				●				●		●	
Marguerite	*Chrysanthemum frutescens*	●							●	●	●		●			●			●				●	
(Blue) Marguerite (see *Felicia*)																								
Mexican bottle plant **(see Ponytail)**																								
Mexican firecracker (see Firecracker vine)																								
Mexican flame vine	*Senecio confusus*		●						●	●	●		●			●			●				●	
Mexican foxglove	***Tetranema roseum***		●						●	●			●			●				●			●	
Ming Aralia (see *Polyscias*)																								
Monkey puzzle	*Araucaria araucana*			●					●		●		●				●			●		●		
Monstera	*Monstera* species				●	●		●	●	●	●		●			●	●		●			●		
Moonstones (see Succulents)																								
Morning glory	*Ipomoea* species	●		●					●	●	●		●				●			●			●	
Moses in the cradle	*Rhoeo* species				●				●	●			●				●			●		●	●	
Moss, Clubmoss	*Selaginella* species					●		●	●				●				●			●		●		
Mother-in-law tongue	*Sansevieria* species	●		●	●	●	●	●	●	●	●			●			●			●		●		
Myrtle	*Myrtus communis*	●		●					●	●	●			●		●			●			●		
Nandina, Heavenly bamboo	*Nandina domestica*	●		●					●		●		●				●			●		●		
Narcissus	*Narcissus* varieties			●					●	●	●		●			●			●				●	
Natal plum	*Carissa grandiflora*	●		●					●	●	●		●			●				●		●	●	
Nephthytis	*Syngonium* species				●	●	●	●	●				●			●				●		●		
Nerine	*Nerine, Lycoris* species	●		●					●		●			●		●			●				●	
Nerve plant (see *Fittonia*)																								
Nicotiana, Flowering tobacco	***Nicotiana alata 'Grandiflora'***			●					●	●	●		●				●			●			●	
Norfolk Island pine	***Araucaria heterophylla***				●	●		●	●	●			●				●		●	●		●		
Oleander	*Nerium oleander*	●		●					●	●	●		●				●			●		●	●	
Olive	*Olea europaea*	●		●					●	●	●			●		●	●		●			●		
Onion, Flowering	*Allium* species	●								●	●		●			●			●				●	
Orange (see Citrus)																								
Orchids:	*Acineta* species	●							●				●				●				●		●	
Fox-brush	*Aerides* species	●									●		●			●				●			●	
Comet	*Angraecum* species			●					●				●			●				●			●	
Spider	*Ansellia* species	●							●					●		●					●		●	
Carnival	*Ascocentrum* species	●							●				●				●				●		●	
No common name	*Bifrenaria* species				●	●		●	●				●				●				●		●	
Lady-of-the-night	*Brassavola* species	●							●						●			●		●			●	

COMMON NAME	BOTANICAL NAME	LIGHT REQUIREMENTS						EXPOSURE				WATER REQUIREMENTS				HUMIDITY			TEMPERATURE			MOST ATTRACTIVE FEATURES		
		Direct sun	Winter direct sun	Bright	Moderate	Low	Very low	North	East	West	South	Keep wet at all times	Keep evenly moist	Approach dryness between waterings	Dry out between waterings	Average house	Moist	Very moist	Cool	Average house	Warm	Foliage	Flowers	Fruit
Cattleya	*Cattleya* species	●							●						●			●		●			●	
No common name	*Coelogyne* species			●					●					●		●				●			●	
Cymbidium	*Cymbidium* species	●							●				●			●			●				●	
No common name	*Dendrobium* species			●						●	●			●		●				●			●	
Clamshell, Butterfly	*Epidendrum* species	●		●					●					●				●		●			●	
Punch and Judy	*Gongora* species			●					●					●		●				●			●	
No common name	*Laelia* species	●		●					●					●		●			●				●	
No common name	*Lycaste* species	●							●					●		●			●				●	
Kite	*Masdevallia* species					●		●					●				●		●				●	
No common name	*Maxillaria* species			●					●								●		●				●	
Pansy	*Miltonia* species			●					●				●			●			●				●	
No common name	*Neofinetia* species			●					●				●			●			●				●	
Lily-of-the-valley, Tiger	*Odontoglossum* species		●						●					●				●	●				●	
Dancing lady, Butterfly	*Oncidium* species	●		●					●				●					●		●			●	
Lady slipper	*Paphiopedilum* species				●				●				●				●	●		●			●	
Moth	*Phalaenopsis* species			●					●				●					●		●			●	
Rodriguez	*Rodriguezia* species			●					●				●				●				●		●	
No common name	*Trichocentrum* species			●					●				●			●				●			●	
Vanda	*Vanda* species			●					●				●				●				●		●	
No common name	*Zygopetalum* species			●					●					●		●				●			●	
Oxalis	*Oxalis* species	●		●					●	●	●		●				●			●		●		
Palms:																								
Areca, Butterfly	*Chrysalidocarpus lutescens*			●	●				●				●		●	●				●		●		
Bamboo	*Chamaedorea* species				●	●			●				●		●	●				●		●		
Chinese fan	*Livistona chinensis*			●					●	●			●				●			●		●		
Dwarf date	*Phoenix roebelenii*			●	●				●				●			●				●		●		
European fan	*Chamaerops humilis*	●		●					●	●	●		●			●			●			●		
Fishtail	*Caryota* species			●					●		●	●					●				●	●		
Kentia, Paradise, Sentry	*Howeia* species			●	●	●		●	●	●	●		●			●				●		●		
Lady	*Rhapis* species				●			●		●			●			●			●			●		
Manila	*Veitchia merrillii*			●					●		●	●					●			●	●	●		
Parlor, Neanthe bella, Dwarf mountain	*Chamaedorea elegans*					●	●	●	●						●					●		●		
Tufted or clustered fishtail	*Caryota mitis*				●			●	●			●					●				●	●		
Pan American friendship plant, Panamiga	*Pilea involucrata*				●			●	●				●			●	●			●		●		
Pandanus	*Pandanus* species				●	●		●	●				●				●	●		●		●		
Papyrus	*Cyperus papyrus*	●		●					●	●	●	●					●			●		●		
Parlor ivy, German ivy	*Senecio mikanioides*		●						●	●	●		●				●		●			●	●	
Parrot flower	*Heliconia psittacorum*	●							●	●	●		●				●				●		●	
Parsley Aralia (see *Polyscias*)																								
Passion flower	*Passiflora* species	●		●						●	●		●				●			●			●	
Peacock plant	*Calathea* species			●	●			●	●	●			●				●				●	●		
Pedilanthus (see Devil's backbone)																								
Peperomia	*Peperomia* species				●			●	●	●					●		●			●		●		
Pepper	*Capsicum* species	●		●				●			●		●				●				●	●		●
Philodendron	*Philodendron* species				●	●	●	●	●	●			●			●	●			●		●		
Piggyback plant	*Tolmiea menziesii*			●	●	●			●	●			●				●			●		●		
Pilea (see Aluminum plant, creeping charlie, Pan American friendship)																								
Pineapple lily	*Eucomis* species			●					●	●	●		●				●			●			●	
Pittosporum (Japanese)	*Pittosporum tobira*	●		●					●	●	●		●				●		●			●	●	
Pleomele	*Pleomele* species			●	●	●		●	●	●		●					●			●	●			

COMMON NAME	BOTANICAL NAME	LIGHT REQUIREMENTS						EXPOSURE				WATER REQUIRE-MENTS				HUMIDITY			TEMPERA-TURE			MOST ATTRAC-TIVE FEATURES		
		Direct sun	Winter direct sun	Bright	Moderate	Low	Very low	North	East	West	South	Keep wet at all times	Keep evenly moist	Approach dryness between waterings	Dry out between waterings	Average house	Moist	Very moist	Cool	Average house	Warm	Foliage	Flowers	Fruit
Plumbago	*Plumbago indica*	•		•					•	•	•		•				•			•		•	•	
Pocketbook plant (see *Calceolaria*)																								
Podocarpus, Japanese Yew	*Podocarpus* species		•	•					•		•		•				•			•		•		
Polka dot plant (see *Hypoestes*)																								
Polyscias	*Polyscias* species		•	•	•				•	•	•		•				•	•	•	•		•		
Pomegranate	*Punica granatum*	•		•						•	•		•				•			•				•
Ponytail	*Beaucarnea recurvata*	•		•	•				•	•	•			•		•				•		•		
Pothos, Devil's ivy	*Scindapsus* species			•	•	•	•	•	•				•				•			•		•		
Powder puff	*Calliandra* species	•		•					•	•	•		•				•				•		•	
Prayer plant	*Maranta* species				•			•	•				•				•			•		•		
Privet (Wax-leafed)	*Ligustrum japonicum*	•		•					•	•	•		•				•			•		•		
Purple Heart	*Setcreasea purpurea*			•					•		•			•			•			•		•		
Queen's wreath	*Petrea volubilis*	•							•		•	•	•				•			•			•	
Rainbow flower (see *Achimenes*)																								
Rain lily	*Zephyranthes* species	•		•					•			•				•			•				•	
Rice-paper plant	*Tetrapanax papyriferus*	•		•					•	•	•		•				•			•		•		
Rosary vine	*Ceropegia* species	•		•					•	•	•				•		•			•		•		
Rose (Miniature)	*Rosa* species	•		•					•	•	•		•				•			•			•	
Rosemary	*Rosmarinus officinalis*	•		•					•	•	•			•			•			•		•	•	
Rubber plant (see Figs)																								
Ruellia	*Ruellia* species			•					•	•			•				•			•			•	
Sage (Blue)	*Eranthemum pulchellum*				•				•		•		•				•				•		•	
Sage (Scarlet)	*Salvia* species	•		•					•	•	•		•				•			•			•	
Sago palm (see *Cycas*)																								
Sapphire flower	*Browallia speciosa 'Major'*	•							•	•	•		•			•				•			•	
Scarborough-lily	*Vallota speciosa*	•		•					•		•		•				•			•			•	
Schefflera	*Brassaia* species			•					•	•	•			•			•			•		•		
Screw pine (see *Pandanus*)																								
Sea grape	*Coccoloba uvifera*			•					•		•		•			•				•		•		
Sedum (see Succulents)																								
Seersucker plant	*Geogenanthus undatus*			•					•		•		•				•				•	•		
Shower tree	*Cassia* species			•					•		•		•			•				•		•	•	
Shrimp plant (Beloperone)	*Justicia* species	•		•					•	•	•				•		•			•		•	•	
Silk oak	*Grevillea robusta*	•		•					•	•	•			•			•			•		•		
Singapore holly	*Malpighia coccigera*	•		•					•		•			•			•			•			•	
Slipperwort (see *Calceolaria*)																								
Snake plant (see Mother-in-law tongue)																								
Society garlic	*Tulbaghia violacea*			•					•		•		•				•			•		•	•	
Spanish shawl	*Heterocentron elegans*	•		•					•	•	•		•				•			•			•	
Sparmannia (see African hemp)																								
Spathe flower	*Spathiphyllum* species			•	•	•		•	•				•				•				•	•	•	
Spider aralia (see Dizygotheca)																								
Spiderplant, Airplane plant	*Chlorophytum* species			•	•			•	•	•	•		•				•			•		•		
Squills	*Scilla* species	•		•					•		•		•			•			•				•	
Star-jasmine	*Trachelospermum jasminoides*		•						•		•			•			•			•		•	•	
Star-of-Bethlehem	*Ornithogalum* species	•		•					•	•	•			•		•			•				•	

Miniature tulips (Tulipa kaufmaniana 'Gold Coin') *can be forced indoors for early blooms.*

COMMON NAME	BOTANICAL NAME	LIGHT REQUIREMENTS						EXPOSURE				WATER REQUIRE-MENTS				HUMIDITY			TEMPERA-TURE			MOST ATTRAC-TIVE FEATURES		
		Direct sun	Winter direct sun	Bright	Moderate	Low	Very low	North	East	West	South	Keep wet at all times	Keep evenly moist	Approach dryness between waterings	Dry out between waterings	Average house	Moist	Very moist	Cool	Average house	Warm	Foliage	Flowers	Fruit
Stephanotis	*Stephanotis floribunda*	●		●					●		●		●				●			●			●	
Stepladder plant	*Costus malorlieanus*			●	●				●	●			●				●				●	●		
Strawberry geranium	*Saxifraga stolonifera*			●	●			●	●				●	●			●		●			●		
Streptosolen	*Streptosolen jamesonii*		●	●					●		●		●				●			●		●	●	
String-of-pearls	*Senecio rowleyanus*			●					●		●		●				●		●			●		
Succulents:																								
Century plant	*Agave* species	●								●	●				●	●				●		●		
No common name	*Aloe* species	●		●						●	●		●			●				●		●	●	
Ice plant	*Aptenia* species	●								●	●		●			●				●			●	
Climbing onion	*Bowiea* species			●	●				●	●	●		●			●			●			●		
Propeller/Rattlesnake/ Scarlet paintbrush	*Crassula* species	●		●					●	●	●		●	●		●				●		●		
No common name	*Echeveria* species	●		●					●	●	●				●	●				●		●	●	
Poinsettia/Crown of thorns	*Euphorbia* species	●		●					●	●	●			●	●	●				●			●	
Tiger jaws	*Faucaria* species	●								●	●				●	●				●		●		
Baby toes	*Fenestraria* species	●		●						●	●				●	●				●		●		
Ox-tongue	*Gasteria* species			●	●				●	●			●			●				●		●		
Zebra/Wart	*Haworthia* species	●		●					●	●	●				●	●				●		●		
No common name	*Kalanchoe* species	●		●					●	●	●				●	●			●	●		●	●	
Living stones/Stone face	*Lithops* species	●								●	●		●			●				●		●	●	
Moonstones	*Pachyphytum* species	●		●					●	●	●				●	●				●		●	●	
Donkey's tail/Coral beads	*Sedum* species	●		●					●	●	●				●	●				●		●		
Starfish flower	*Stapelia* species	●		●					●	●	●		●			●				●			●	
Swedish ivy	*Plectranthus* species			●	●				●	●	●		●				●			●		●		
Sweet olive	*Osmanthus fragrans*	●		●					●	●	●		●				●		●			●		
Syngonium (see *Nephthytis*)																								
Taffeta plant	*Hoffmannia* species				●			●	●				●				●					●		
Tangerine (see Citrus)																								
Temple bells	*Smithiantha* species			●	●				●		●		●				●				●		●	
Ti tree (see Hawaiian ti)																								
Treevine, Begonia Cissus	*Cissus discolor*			●	●				●		●		●				●			●			●	
Tree ivy (see *Fatshedera*)																								
Tulips	*Tulipa* species			●					●	●	●		●			●			●				●	
Turk's cap	*Malvaviscus arboreus*	●		●					●	●	●		●				●		●				●	
Umbrella plant (see *Cyperus*)																								
Umbrella tree (see Schefflera)																								
Vegetables (See Annuals)																								
Veltheimia	*Veltheimia* species			●					●				●			●			●			●		
Velvet plant	*Gynura aurantiaca*	●		●					●	●			●				●				●	●		
Velvet plant (Trailing) (see *Ruellia*)																								
Wandering Jew, Inch plant:	*Setcreasea* species		●	●	●			●	●					●		●	●		●			●		
	Tradescantia species			●	●			●	●					●		●			●			●		
	Zebrina species			●	●			●	●				●				●			●		●		
Wand flower	*Sparaxis* species	●		●					●	●	●		●			●			●				●	
Wax plant	*Hoya* species	●		●	●				●	●				●			●			●		●	●	
Winter creeper	*Euonymus radicans*			●					●		●		●			●			●			●		
Wire plant	*Muehlenbeckia* species	●		●					●	●			●				●			●		●		
Yew (Japanese) (see *Podocarpus*)																								
Yucca	*Yucca* species	●		●					●	●	●				●		●			●		●	●	
Zebra plant	*Aphelandra* species			●					●	●			●			●				●		●	●	
Zephyr lily	*Zephyranthes* species	●		●					●	●	●						●			●			●	

Delicate white blooms accent the variegated foliage of a wandering Jew (Tradescantia albiflora) ⇨

Charles Spencer

Let the light shine in!

How to get more (or less) sunlight into your home. Easy-to-grow plants for all exposures. Light and types of "windows."

The amount and exposure of the natural sunlight available must become one of the indoor gardeners' primary considerations for successfully maintaining healthy plants. Some plants such as cacti and succulents can take direct light all day. Some can take direct sun for only short periods while others can take filtered light only or they will sunburn. Some plants tolerate only indirect light. Ordinary window glass traps heat, so some plants that thrive in direct sun outdoors may not be able to take it indoors.

As the sun performs its daily ritual the light conditions in the home are constantly changing. A spot that was in shadow moments ago is now in full light as the sun moves through the sky and another area that was in filtered sunlight is now in shadow. These fluid conditions have an enormous impact on plants. Yet most of us are only vaguely aware of the changing intensity and direction of sunlight at various times of the day. A little light-awareness will help! Consider the following Three Act Play which describes the sun's interplay with various household exposures.

Artist Lenny Meyer has created a relaxing oasis in a north light bay window. Clear glass plates catch drainage from hanging terracotta pots. The kangaroo vine (Cissus antarctica) *on left wall has attached its tendrils to white string trellis.*

Changing light: the three act play

In our common everyday experience, the sun rises, the sun sets. In between it creates an ever-changing setting of light patterns and intensities in the home. The first act of this daily performance opens with the first glow in the eastern sky. The dark room melts into shadows and these in turn give way to soft forms. The light is weak, tentative.

This glow ends suddenly with the orangey glint of sunlight stealing over the eastern horizon. More glimmer than real light, the sun's first effort streams across the room and shimmers on the far wall. There is a pause as the sun pops above the horizon; then movement as the zebra-stripe pattern of light, through the venetian blinds, moves down the wall and across the floor. Morning is here, with it all the quiet radiation from our central solar body.

The eastern light is warm, yet cool—the coolest of the three exposures receiving direct sunlight. As the morning progresses the brightness of the light increases. It is the opening act on a normal day—daily, that is, when clouds don't interrupt or cancel the performance!

The second act commences when the sun abandons the eastern exposure and demands attention at centerstage—on the south. The light is brighter now because the sun's rays have less atmosphere to penetrate before striking Earth; the direct sunlight retreats out of the room.

At mid-day the light is at its most intense because the sun is closest to us then and has the least atmosphere to filter through. The light has reached its brightest about noon—depending on where you are located relative to the center of your time zone. Outdoors, shadows form strong contrast to the brilliant light—this bold, dazzling, even harsh noonday light. Indoors, however, in summer at mid-day, the rooms of the house shielded by the roof and its overhangs, receive no direct sunlight.

Midafternoon opens the final sequence in the performance. The sun loses its upward thrust and begins to describe its downward arc. The light, still intense, begins to move back into the room. The sun has migrated—this time from the south side to the western exposure. The light is still bright—and, as the day progresses, completely fills the room again. Spots that had been in shadow are now in direct sunlight and the heat rays shimmer on a west-facing windowsill. In the late afternoon, light color changes subtly from sparkling white to cream to golden yellow as the intensity drops off. Shadows lengthen and the hint of restfulness is felt.

The sun sinks slowly below the horizon and its rays reach deep into the room. Sheer curtains are a golden yellow as it lowers ever so slowly across the western horizon. The half-light of dusk brings the room into everincreasing darkness and our Three Act Play is concluded.

Notes for "window" graphs

The following information refers to the chart on the next two pages (44 and 45).

*The light values shown by the **colored bars** represent the light entering a room with a single window wall, from each cardinal orientation, on the day of the summer solstice. The measurements, taken on a treeless hill top at 37° 58′ latitude, represent optimum light.*

The morning and afternoon "humps" on the north window graph are caused by the low angle sunlight entering the room beneath the overhang: the reflected light from the side walls was sufficient to double the f.c. reading. (The meter was not in this direct light.) This effect would be less pronounced in a room with smaller windows. With less overhang a similar plateau would appear in the center portion of the south window graph caused by the reflected light from sun shining on the floor.

*The light values shown by the **white line** are measurements made in the east window shown in the photographs. Light entrance into this window is delayed by a wooded hill to the east, and periodically obstructed by tree trunks and high open foliage. The area of open unobscured sky, seen from the plant position in the window, is considerably less than 50% of that which would be present without the hill and the trees; as seen on the graph the afternoon light is approximately ¼ the intensity that could be expected if this window received "full" skylight. The window still receives about three times the radiant energy that a "good" north window gets.*

*It should be noted that the light intensity scale on the graph is a geometric progression, each ascending increment is double the one below it. While the tallest bars, representing direct sunlight are approximately double the height of the bars representing indirect or skylight, the full sunlight intensity (8000 f.c.) is about 16 times the "skylight" intensity (500 f.c.). **It is useful to realize that 4 minutes of direct sunlight provides radiant energy equal to about an hour of full skylight.** In the window photographed, the direct morning sun provides about 15 times as much of the total light as the indirect sky provides.*

A. *7:00 a.m. The sun is already up on the other side of the hill to the east, but not in this window.*

B. *8:15 a.m. The first direct sun hits the sago palm in the window.*

C. *10:15 a.m. Tree shadows cross the window: filtered light through a high open tree canopy.*

Light values at fixed intervals by exposures

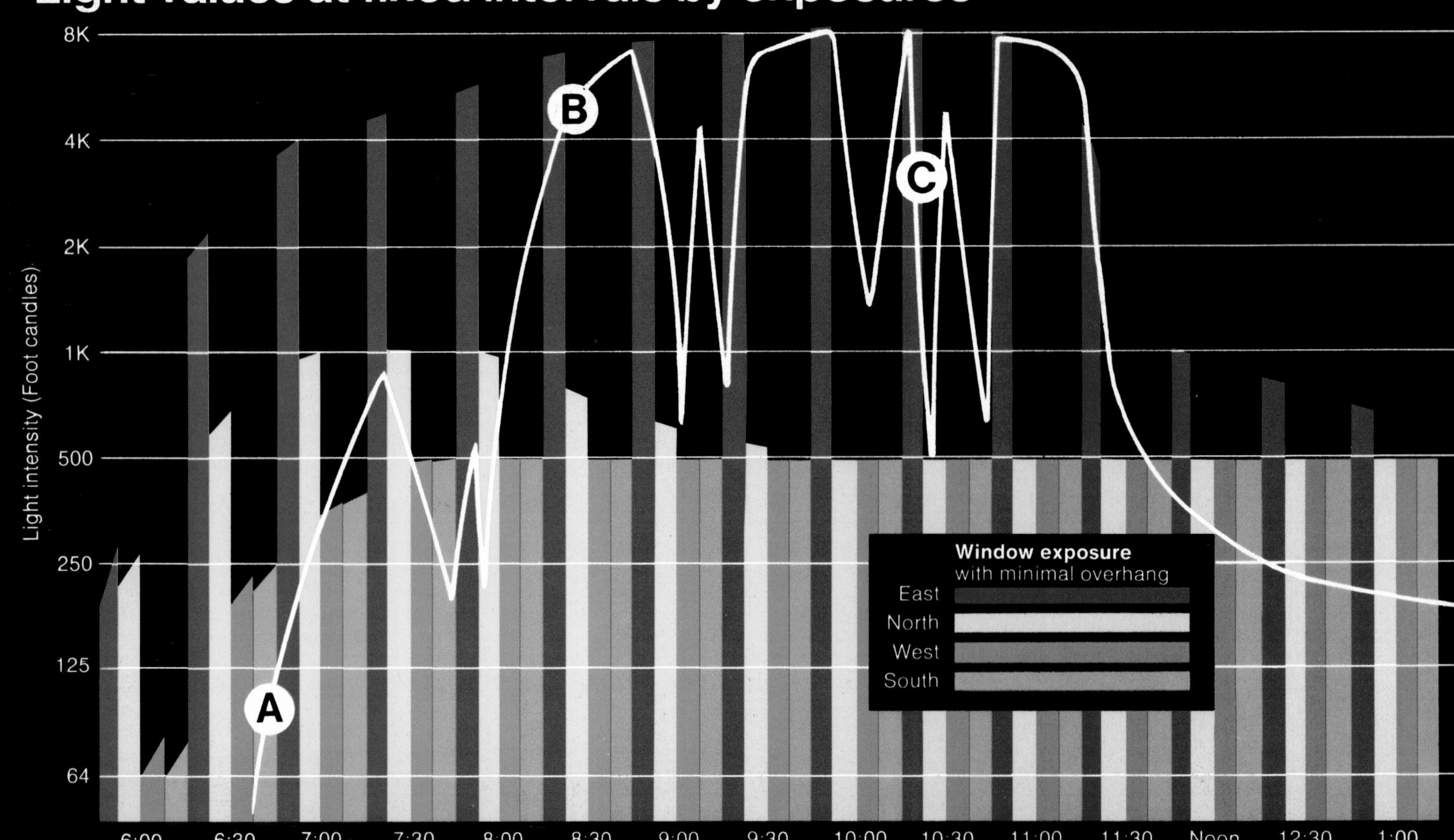

D. *2:30 p.m. Strong afternoon light silhouettes the plants; light on the palm comes largely from the sky all afternoon.*

E. *6:00 p.m. Now the light on the hill is softer, and we can see skylight glisten on the palm fronds.*

F. *7:15 p.m. A last late bit of direct light through a northwestern window passes briefly across the plants.*

D E F

1:30 2:00 2:30 3:00 3:30 4:00 4:30 5:00 5:30 6:00 6:30 7:00 7:30 8:00 8:30

The cool North

The United States, being in the Northern Hemisphere, always receives its sunlight from the south. Even at summer solstice (June 21), when the sun is as far north as it comes, it is just far enough north to peek into a north window! Consequently, of the four exposures, the northern exposure receives the least light and the least heat year-round. Northern light is fairly constant throughout the day and there is less footcandle variation than there is with the other three exposures.

Because of the low light in north windows, maintaining healthy plants can be a challenge. A northern windowsill can register as low as 200 f.c. on a clear mid-winter day. There are, however, a wide variety of low-light tolerant foliage plants suitable for this exposure.

The diagram below and those on the following pages are based on:

- a winter noonday;
- a clear atmosphere and a strong sun;
- a 38°-42° belt of latitude across the United States which includes Washington, Philadelphia, New York, Boston, Detroit, Chicago, St. Louis, Denver, Salt Lake City and San Francisco.

The values shown would be higher in other seasons and for those areas south of latitude 38°. They would be lower at any other time of day, or north of latitude 42°.

The dim-moderate-bright values are based on the following f.c. ranges:

- 50-200 dim light;
- 200-500 moderate light;
- 500-1000 bright light.

They are the categories used to describe minimum lighting requirements for plants in charts on pages 28-41.

To generalize for the sake of brevity, plants grown especially for their green foliage will tolerate north exposure even if they prefer brighter light. Colored foliage plants frequently prefer somewhat brighter light, losing some of their color with a north exposure. Plants grown for their flowers usually need brighter exposure than north. However, you will find exceptions to this generalization. If any plants given a north exposure grow leggy, move them to an east exposure (or supplement with fluorescent plant lamp).

Plants for a northern exposure

Refer to the Plant Selection Guide beginning on page 28 for specific culture and more unusual plants.

Key to code: **(B)** Bright **(H)** High humidity

Asparagus fern (*Asparagus* species)

Baby's tears *(Helxine soleirolii)* **(B-H)**

Caladium species

Cast iron plant *(Aspidistra elatior)*

Chinese evergreen *(Aglaonema* species)

Cobra lily *(Darlingtonia californica)*

Creeping charlie *(Pilea nummulariifolia)* **(H)**

Creeping fig *(Ficus pumila)* **(B-H)**

Dracaena species

Dumb cane *(Dieffenbachia* species)

False aralia *(Dizygotheca elegantissima)* **(B-H)**

Ferns: *Adiantum, Asplenium, Cyrtomium, Davallia, Nephrolepis, Phyllitis* and *Pteris* species **(H)**

Flagplant *(Acorus gramineus)* **(H)**

Goldust plant *(Aucuba japonica 'Variegata')*

Grape ivy *(Cissus rhombifolia)*

Kangaroo vine *(Cissus antarctica)*

Leopard plant *(Ligularia tussilaginea)*

Mistletoe cactus *(Rhipsalis* species) **(B-H)**

Monstera species **(H)**

Moss *(Selaginella* species) **(H)**

Mother-in-law tongue *(Sansevieria* species)

Nephthytis *(Syngonium* species)

Nerve plant *(Fittonia* species) **(H)**

Norfolk Island pine *(Araucaria heterophylla)* **(B)**

Orchids: *Bifrenaria* and *Masdevallia* species **(H)**

Palms: *Caryota, Chamaedorea, Howeia* and *Rhapis* **(B)** species

Panamiga *(Pilea involucrata)* **(H)**

Peperomia species **(B)**

Philodendron species

Pleomele species

Pothos *(Scindapsus* species)

Prayer plant *(Maranta* species) **(H)**

Rubber tree *(Ficus elastica)*

Sago palm *(Cycas* species)

Screw pine *(Pandanus* species) **(B)**

Spathe flower *(Spathiphyllum* species) **(H)**

Spider plant *(Chlorophytum* species)

Strawberry geranium *(Saxifraga stolonifera)* **(B)**

Wandering Jew *(Setcreasea, Tradescantia* and *Zebrina* species) **(B)**

North exposure

Positioning of plants for optimum light requirement in relation to a 3 ft. x 5 ft. window

Glass | Glass | Glass

1' 2' 3' 4' 5' 6' 7' 8' 9' 10'

3' 2' 1' 1' 2' 3' 2' 1' 1' 2' 3' 2' 1' 1' 2' 3'

Numbers on grid indicate distance from window in feet

BRIGHT: Glass obstructed / Not obstructed

MODERATE: Glass obstructed / Not obstructed

DIM: Glass obstructed / Not obstructed

Plant identification (opposite page):

a. Spathiphyllum 'Clevelandii' *(White flag)*
b. Cissus rhombifolia *(Grape ivy)*
c. Adiantum *(Maidenhair fern)*
d. Aglaonema *(Chinese evergreen)*
e. Philodendron panduriforme *(Fiddle-leaf Philodendron)*
f. Dracaena massangeana
g. Chamaedorea elegans *(Dwarf parlor palm)*
h. Scindapsus aureus *(Golden pothos)*
i. Dieffenbachia amoena *(Dumb cane)*
j. Ficus pumila *(Creeping fig)*

a.
b.
c.
d.
e.
g.
f.
h.
j.
i.

a.
b.
c.
d.
e.
h.
g.
j.
f.
i.
l.
m.
k.

The exotic East

The eastern exposure receives direct morning light from sunrise to near mid-day. Footcandle readings can reach 5000-8000. As the morning progresses, the direct sun recedes from the room. This is hastened if there are eaves or other outside overhangs. The east room is cooler than southern or western exposures because the house has absorbed less radiant heat. Most plants will take some direct sun from an eastern exposure, particularly early in the morning.

Most plants will tolerate an east exposure; a preponderance of those normally grown indoors actually prefer it. It is cooler than either south or west, therefore less dehydrating. The early rays of the sun act as an alarm clock, awakening the plants to their day's task—photosynthesis. Many of the most successful gardeners automatically try an east exposure for any plant about which light preference is in question. This list, necessarily, includes only the most available plants for indoor gardeners, but could be expanded tenfold if space permitted.

Plants for an eastern exposure

Refer to the Plant Selection Guide beginning on page 28 for specific culture and more unusual plants.

Key to code: **(B)** Bright **(F)** Filtered sun **(H)** High humidity **(C)** Cool

African violets *(Saintpaulia* species)
Aluminum plant *(Pilea cadierei)* **(H)**
Anthurium species
Aralia, Japanese *(Fatsia japonica)* **(C)**
Artillery plant *(Pilea microphylla)* **(H)**
Asparagus fern *(Asparagus* species)
Avocado *(Persea americana)* **(B-H)**
Azalea *(Rhododendron* species) **(B-H)**
Baby's tears *(Helxine soleirolii)* **(H)**
Begonia species **(B-H-C)**
Bromeliads (many species) **(H)**
Calla lily *(Zantedeschia* species) **(B**
Cast iron plant *(Aspidistra elatior)*
Chenille plant *(Acalypha hispida)*
Chinese evergreen *(Aglaonema* species)
Chinese lemon *(Citrus limon 'Meyer')* **(B)**
Christmas cactus *(Schlumbergera* species) **(B-H)**
Clerodendrum species
Coleus species
Columnea species
Creeping charlie *(Pilea nummulariifolia)* **(H)**
Cyperus species **(C)**
Devil's backbone *(Pedilanthus tithymaloides)*
Dieffenbachia species
Dracaena species
False aralia *(Dizygotheca elegantissima)* **(H)**
Ferns (many species) **(H-F-C)**
Ficus species
Flame violet *(Episcia* species) **(H)**
Flowering maple *(Abutilon* species) **(B)**
Fuchsia species **(B-H-C)**
Geranium *(Pelargonium* species) **(B)**
Ginger (many species) **(B-H)**
Gloxinia *(Sinningia* species)
Goldfish *(Nematanthus* species) **(H)**
Grape ivy *(Cissus rhombifolia)*
Hawaiian ti *(Cordyline terminalis)*
Hibiscus rosa-sinensis **(B-H)**
Ivy *(Hedera* species) **(C)**
Kangaroo vine *(Cissus antarctica)*
Lipstick vine *(Aeschynanthus* species)
Ming aralia *(Polyscias* species) **(B-H)**
Mistletoe cactus *(Rhipsalis* species) **(H)**
Moses-in-the-cradle *(Rhoeo* species) **(C)**
Moss *(Selaginella* species) **(H-F)**
Mother-in-law tongue *(Sansevieria)*
Myrtle *(Myrtus communis)* **(B)**
Nephthytis *(Syngonium* species)
Nerve plant *(Fittonia* species) **(H)**
Norfolk Island pine *(Araucaria heterophylla)*
Orchid cactus *(Ephiphyllum* species)
Orchids (many species) **(B-H)**
Palms (many species) **(B-C)**
Papyrus *(Cyperus papyrus)* **(C-M)**
Peperomia species **(H)**
Philodendron species
Piggyback *(Tolmiea menziesii)*
Pink polkadot *(Hypoestes* species)
Pleomele species
Pocketbook *(Calceolaria* species) **(C)**
Ponytail *(Beaucarnea recurvata)* **(B)**
Pothos *(Scindapsus* species)
Prayer plant *(Maranta* species) **(H)**
Schefflera *(Brassaia* species)
Spathe flower *(Spathiphyllum* species)
Spider plant *(Chlorophytum* species) **(H-C)**
String-of-pearls *(Senecio rowleyanus)*
Succulents (many species)
Swedish ivy *(Plectranthus* species) **(H-C)**
Tree ivy *(Fatshedera lizei)* **(H)**
Velvet plant *(Gynura aurantiaca)* **(B-H)**
Wandering Jew *(Setcreasea, Tradescantia* and *Zebrina* species) **(B-H-C)**
Wax plant *(Hoya* species) **(B-H)**
Zebra plant *(Aphelandra* species)

Plant identification (opposite page):

a. Dizygotheca elegantissima *(False aralia)*
b. Nephrolepis exaltata 'Bostoniensis' *(Boston fern)*
c. Ficus benjamina *(Weeping fig)*
d. Pleomele reflexa
e. Begonia rex cultivar *(Rex begonia)*
f. Dryopteris *(Fern)*
g. Cordyline terminalis *(Hawaiian ti)*
h. Tolmiea menziesii *(Piggyback)*
i. Saintpaulia ionantha *(African violet)*
j. Polyscias *(Ming aralia)*
k. Aeschynanthus lobbianus *(Lipstick vine)*
l. Sinningia speciosa *(Gloxinia)*
m. Tradescantia *(Wandering Jew)*

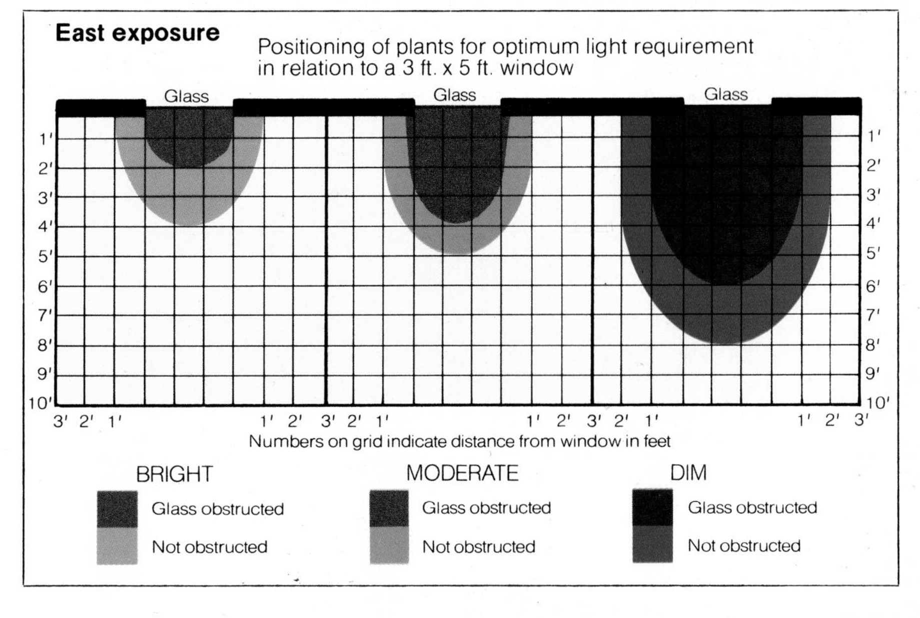

a.
b.
c.
d.
e.
f.
g.
h.
i.
j.
k.
l.

The sunny South

The seasonal variation in southern light is greater than any other exposure. Fortunately for the indoor gardener, the lower winter sun streams across the room for most of the daylight hours. On a bright, sunny winter day the southern room is your greenhouse. In the summer, when the sun is farther north than in winter, the sun rises at a sharp angle in the morning and is high in the sky by noon, consequently there is direct sunlight only immediately in front of a southern window at mid-day. In fact, if there is a wide overhang of the eaves outside, the sun may not enter the room at all.

The amount of light that enters a home through a southern window will be only a portion of the available light outdoors on a clear day. The outdoor sun at high noon on a summer day may register as much as 10,000 f.c.—or more if there is reflected light off buildings as well. Indoors, however, on the same sunny day, a southern window with its wide overhang of eaves outside, will receive about the same amount of light as those in a northern exposure.

South and west exposures are interchangeable for most plants, according to summer-winter changes in the sun's position. Diffusing direct light from the winter south exposure, applying additional humidity (with perhaps some cooling) makes it possible to place many more plants than this list indicates in a south facing area. In fact, with cooling and humidity, all but those with a definite north preference may be grown in a south winter exposure. In the summer, most on the east exposure list will be very happy with a south exposure, and those on the west exposure list requiring filtered west light will thrive with the south exposure in summer.

Plants for a southern exposure

Refer to the Plant Selection Guide beginning on page 28 for specific culture and more unusual plants.

Key to code: (H) High humidity (C) Cool (F) Filtered sun

Amaryllis *(Hippeastrum vittatum)*
Annuals and vegetables (many species)
Aralia, Japanese *(Fatsia japonica)*
Asparagus fern *(Asparagus* species)
Avocado *(Persea americana)* (H)
Azalea *(Rhododendron* species) (H-C)
Bamboo *(Bambusa* species) (H)
Banana *(Musa* species)
Begonia (fibrous-rooted and wax, *B. semperflorens)* (H-F)
Bleeding heart *(Clerodendrum* species)
Bromeliads *(Ananas, Catopsis* and *Guzmania* species)
Cacti (many species)
Calla lily *(Zantedeschia* species)
Camellia species
Chenille plant *(Acalypha* species)
Citrus species
Coffee *(Coffea* species) (H)
Coleus species
Double decker plant *(Sinningia verticillata)*
Elephant ears *(Colocasia* species)
Ferns *(Alsophila, Cyathea* and *Polypodium species)* (H)
Flame violet (*Episcia* species) (H)
Flame-of-the-woods *(Ixora* species)
Flowering tobacco *(Nicotiana alata grandiflora)*
Gardenia jasminoides (H)
Geranium *(Pelargonium* species)
Heavenly bamboo *(Nandina domestica)*
Hibiscus rosa-sinensis
Ivy *(Hedera* species)
Kaffir lily *(Clivia* species)
Lily-of-the Nile *(Agapanthus* species)
Lipstick vine *(Aeschynanthus* species)
Myrtle *(Myrtus communis)*
Natal plum *(Carissa grandiflora)*
Pink polkadot *(Hypoestes* species) (H)
Olive *(Olea europaea)*
Orchids *(Aerides* and *Dendrobium* species)
Oxalis species (H)
Palms *(Caryota, Chamaerops, Howeia* and *Veitchia* species)
Papyrus *(Cyperus papyrus)* (H-F)
Parlor ivy *(Senecio mikanioides)* (C)
Passion flower *(Passiflora* species) (H)
Pepper *(Capsicum* species)
Pittosporum tobira
Podocarpus species
Ponytail *(Beaucarnea recurvata)*
Privet *(Ligustrum japonicum)*
Rainbow flower *(Achimenes* species)
Schefflera *(Brassaia* species) (H)
Shrimp plant *(Justicia* species) (H)
Singapore holly *(Malpighia coccigera)*
Spider plant *(Chlorophytum* species) (F)
String-of-pearls *(Senecio rowleyanus)* (H)
Succulents (many species)
Swedish ivy *(Plectranthus* species) (H)
Sweet olive *(Osmanthus fragrans)* (H-C)
Treevine *(Cissus discolor)*
Winter creeper *(Euonymus radicans)*
Zephyr lily *(Zephyranthes* species)

Plant identification (opposite page):

- ***a.*** Beaucarnea recurvata *(Ponytail)*
- ***b.*** Citrus mitis *(Calamondin orange)*
- ***c.*** Echeveria
- ***d.*** Nandina domestica *(Heavenly bamboo)*
- ***e.*** Echeveria affinis
- ***f.*** Coleus blumei
- ***g.*** Coleus blumei
- ***h.*** Rhododendron *species (Azalea)*
- ***i.*** Coffea arabica *(Coffee)*
- ***j.*** Sedum morganianum *(Donkey's tail)*
- ***k.*** Kalanchoe blossfeldiana
- ***l.*** Echinopsis multiplex *(Barrel cactus)*

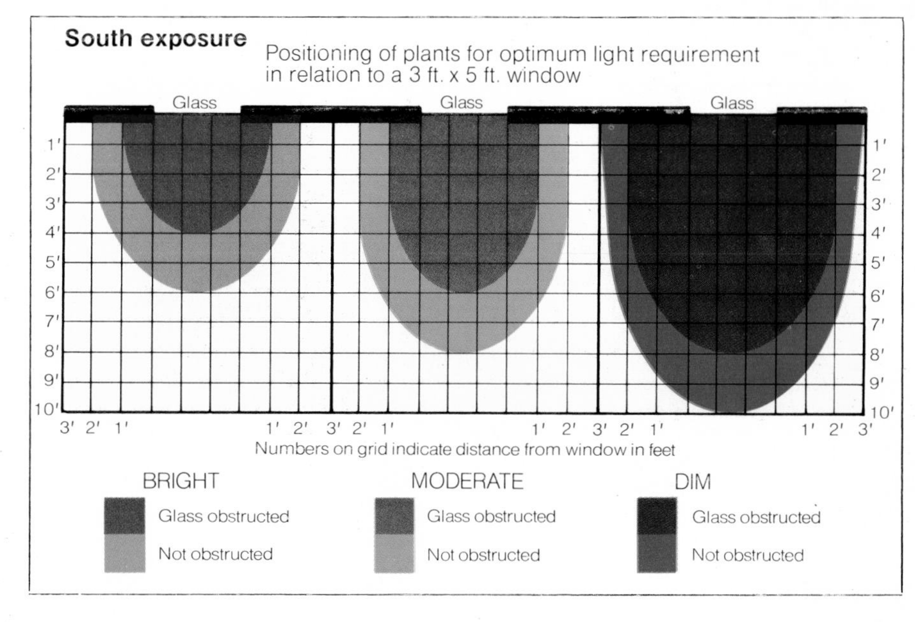

The wild West

Western exposures have the highest summer temperatures. Photosynthesis does not occur at temperatures above 85°. Sustained heat in a western window will kill many plants.

Cacti, succulents and annuals are among the few plant types that can usually take a hot windowsill environment on the western side. For protection, other plants should be moved two or three feet back into the room where less intense late afternoon rays will do little harm. This exposure receives about the same amount of light as the eastern exposure—about four hours of direct sunlight a day.

This poor indoor garden climate, however, is easy to overcome and can be turned into an ideal growing area for more plants than you could possibly accommodate under your roof! Three provisions are essential: (a) Humidity. (Moist air counteracts heat dehydration) (b) Good air circulation. (Air movement is necessary, even for cacti) (c) Curtaining. (It filters out the sun's scorching rays.) From November to March, for any plants requiring bright light, it is not necessary to use curtaining with a west exposure.

Not only does light vary by exposures and by season, it varies by elevation and latitude. Light intensity is much higher in the mountains than at sea level—not only because places at higher elevations are closer to the sun, but because the atmosphere is thinner there, and less light is filtered out before striking Earth.

The indirect light through a window may be less intense at these elevations, since the deeper blue sky reflects less light.

Air conditioning, while providing some cool air movement, is very drying; so with air conditioning, give special attention to humidity.

Plants for a western exposure

Refer to the Plant Selection Guide beginning on page 28 for specific culture and more unusual plants.

Key to code:

(B) Bright **(H)** High humidity
(C) Cool **(F)** Filtered sun

African violets *(Saintpaulia* species) **(H-F-C)**
Algerian ivy *(Hedera canariensis)* **(F-C)**
Aluminum plant *(Pilea cadierei)* **(H)**
Amaryllis *(Hippeastrum vittatum)*
Annuals and vegetables (many species) **(B)**
Aralia *(Fatsia japonica)*
Artillery plant *(Pilea microphylla)* **(H)**
Asparagus fern *(Asparagus* species)
Avocado *(Persea americana)* **(H)**
Bamboo *(Bambusa* species) **(H)**
Banana *(Musa* species)
Begonia species
Bleeding heart *(Clerodendrum* species)
Bromeliads (many species) **(H-C)**
Cacti (many species)
Camellia species
Cast iron plant *(Aspidistra elatior)*
Chenille plant *(Acalypha hispida)*
Citrus species **(C)**
Coffee *(Coffea* species)
Coleus species **(F)**
Columnea species **(H-F-C)**
Creeping charlie *(Pilea nummulariifolia)* **(H-F-C)**
Croton *(Codiaeum* species)
Elephant ears *(Colocasia* species)
Ferns *(Alsophila, Cyathea* and *Platycerium* species) **(H-F-C)**
Ficus species **(H-F)**
Firecracker flower *(Crossandra* species)
Flame violet *(Episcia* species) **(H)**
Flame-of-the-woods *(Ixora* species)
Flowering tobacco *(Nicotiana alata*
Geranium *(Pelargonium* species)
Goldfish *(Nematanthus* species) **(H)**
b. Araucaria heterophylla
Lily of the Nile *(Agapanthus* species)
Ming aralia *(Polyscias* species) **(H-F)**
Mother-in-law tongue *(Sansevieria* species) **(F)**
Myrtle *(Myrtus communis)*
Nerve plant *(Fittonia* species) **(H)**
Orchids *(Dendrobium* species) **(H)**
Oxalis species **(H-C)**
Palms *(Chamaerops, Howeia, Livistona* and *Rhapis* species) **(H-F-C)**
Parlor ivy *(Senecio mikanioides)* **(C)**
Peperomia species **(H-F)**
Pleomele species
Piggyback *(Tolmeia menziesii)*
Pittosporum species
Rainbow flower *(Achimenes* species)
Rosary vine *(Ceropegia* species)
Sago palm *(Cycas* species) **(H)**
Schefflera *(Brassaia* species)
Spider plant *(Chlorophytum* species) **(H-F-C)**
Succulents (many species)
Swedish ivy *(Plectranthus* species) **(H-F-C)**
Tree ivy *(Fatshedera lizei)* **(H-C)**
Velvet plant *(Gynura aurantiaca)* **(H-F)**
Wax plant *(Hoya* species)
Zebra plant *(Aphelandra* species)

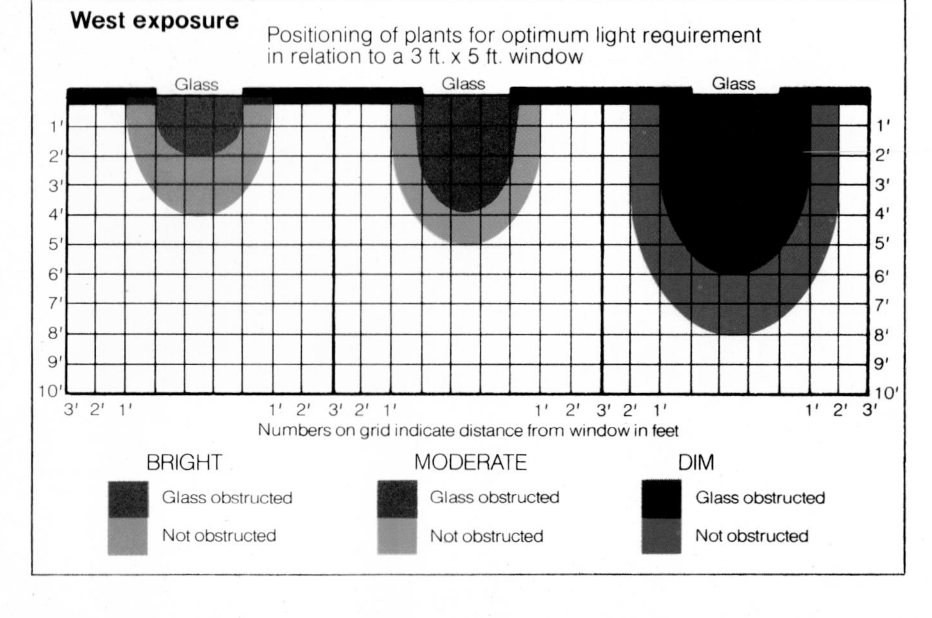

Plant identification (opposite page):

a. Cissus antarctica *(Kangaroo vine)*
b. Araucaria heterophylla *(Norfolk Island pine)*
c. Pilea microphylla *(Artillery fern)*
d. Lilium tigrinum *(Tiger lily)*
e. Schefflera venulosa
f. Codiaeum *species (Croton)*
g. Sansevieria trifasciata *(Mother-in-law's tongue)*
h. Ficus elastica *(Rubber plant)*
i. Aechmea fasciata *(Living vase)*
j. Fittonia verschaffeltii argyroneura *(Nerve plant)*

a.
b.
c.
d.
e.
f.
g.
h.
i.
j.

Light and types of windows

"Windows" should be thought of as any openings which allow the passage of sunlight into a home from the outside. In this context skylights and open doors are also "windows." Since sunlight is "free" and has the proper spectrum for plant growth, *the major goal of the indoor plant gardener is to get the most use of 'windows' as openings for natural light.* With low cost energy rapidly becoming a thing of the past, it is doubly important to reduce our reliance on artificial light to the greatest possible extent.

The windows with which we are most familiar are the classic double-hung and the aluminum frame crank-out type. Both of these forms are 'cut-out' type windows, usually about three feet off the floor, occupying 20-30 percent of a vertical wall. These admit direct sunlight which covers a comparable 20-30 percent of a room's floor surface. During the day when the sun's rays strike the window glass in a more vertical position there is less sunlit-covered surface. In fact, at noon on a summer day there will be almost no direct sunlight coming through a south window and none through east or west windows. More direct sunlight enters the home in the winter when the sun is lower in the sky. But because the sun is farther away from the Earth, winter light intensities tend to be considerably lower, too.

The glass in your windows may decrease the intensity of the light by as little as 10% or as much as 50% depending on the angle the light enters. The more obliquely the light strikes the glass the more the intensity is decreased: thus the high overhead light at midday in summer coming in a south window may be cut in half, while the early morning and late afternoon direct sun through east and west windows is decreased by about 20% and diffuse skylight by still less. These are of course approximations—a lot depends on the thickness and color of the glass and how clean you keep your windows.

Modern American architecture has broken away from the traditional rectangular shell that has comprised the bulk of our houses and apartments. With the new forms have come new 'windows'—new ways to admit light, new opportunities for more light on interior gardens. Some of these types are explored here for the homeowner who is fortunate enough to enjoy a home with these innovative windows, or who may want to incorporate them into an existing building or into plans for future structures.

CLERESTORY WINDOWS have been used for many years, but perhaps not widely enough, considering their advantages in terms of interior gardening possibilities. They are strip windows, oriented horizontally, high in the wall. They provide better-distributed light than a conventional window of the same surface size because they throw the light farther into the room. This can make for some dramatic, spotlighting effects with large plants. These windows allow flexible architectural treatment because they free up the whole wall below them for other uses and provide a horizontal proportion to the room. Potted plants are attractive in a clerestory window sill, especially trailing varieties like Swedish ivy (Plectranthus australis) or kangaroo vine (Cissus antarctica).

WINDOW WALLS have become more and more popular. Although they make insulation against heat loss difficult, they provide the greatest lighting flexibility of any window type if only because they permit considerably more light to enter than any other kind. Reducing light by means of shades, shutters, curtains, or blinds is certainly easier than trying to *add* 200 or 300 f.c. to a dark corner. One great advantage to window walls is their ability to merge the plantings indoors with those outdoors. This gives an illusion of total unity with the garden. The effect is enhanced two-fold when the same plants are used on both sides of the glass, making the wall visually "disappear." Hanging plants such as donkey's tail *(Sedum morganianum),* Spanish-shawl *Heterocentron elegans),* or spider plant *(Chlorophytum)* can be suspended at various heights, creating a wall of green in the window.

VERTICAL SLIT WINDOWS. Where two walls meet, floor-to-ceiling vertical slit windows can be introduced. Vertical slit windows create a strong indoor-outdoor relationship, reinforced by matching indoor and outdoor plantings. However, in choosing plants that must grow both indoors and outdoors, be sure to select subjects that are adaptable to either area. If the outside area is protected from full sun and harsh wind, possibilities of choice are many. The aralias offer a diversity of foliage patterns and adaptability almost unequaled for this type planting. Among them are the scheffleras, Brassia actinophylla, and Dizygotheca elegantissima. The latter, however, has a definite juvenile foliage and mature foliage. Indoors, it usually maintains its juvenile form,

Hanging baskets fill a livingroom window wall, creating a living curtain.

A series of large windows admit light through most of the day, allowing designer Ron Hildebrand to meet light requirements of a wide range of plants.

but when placed outdoors, the mature foliage develops and the plant loses its narrow, delicate looking leaves, which are replaced by broad umbrellas of beautiful foliage.

BAY OR BOW WINDOWS. A classic window type, these differ from standard double-hung casements in that they protrude beyond the wall plane to gather up more light. A south facing bay window may be in light 12 to 14 hours a day. They are most effective on a north wall where they will be able to introduce a small amount of direct eastern and western sunlight and will therefore expand the planting possibilities. Many plants will be healthier in a northern bay window than in a northern casement.

EXTENDED WINDOWS. Similar in principal to the bay window, extended windows can take various forms, from half greenhouses appended into the living room to a sawtooth extended out to capture needed light. These windows can provide the effect of a mini-conservatory inside the home where plants can be enjoyed more than if they were growing in a detached greenhouse.

SKYLIGHTS. These holes-in-the-ceiling have been around for a long time, employed primarily in industrial settings. The major reason for installation of skylights in the home is usually utilitarian—to bring daylight into inside, windowless rooms that would have no other access to natural light. Plants, if introduced at all are afterthoughts. But because skylights can bring light deep into a home, their light pattern becomes extremely effective in providing life-giving sunlight to areas that otherwise couldn't even support foliage plants. Because skylights do provide direct overhead light, however, a word of caution is necessary. The patch of sun, as it moves across the room, could possibly overwhelm light-sensitive plants, particularly if placed too close to the ceiling (such as a hanging plant or a very tall-growing specimen). A typical 4′ x 4′ skylight should present no problem, since direct light will remain on the area beneath it for only about a half hour. More harm, however, could come from heat generated beneath the skylight than from the actual sunlight itself. A check of the temperature should be made to determine the highest degree of heat to be expected when the sun is at its zenith. (Take the thermometer reading at the same level as the top of the plant or plants involved) Remember—humidity can help to offset the effects of heat, counteracting loss of moisture in the plant.

The design possibilities of skylights are endless. The simple 'bubble' type can be installed singly or in tandem. A whole ceiling could be made into skylights, turning a living room into a veritable greenhouse. Hanging plants can be hung either in the well of the skylight, if there is one, or randomly from exposed beams, creating a garden-in-the-sky.

HIP-ROOF SKYLIGHTS. These are skylights running the length of the house, on one or both sides of the hip. Because of the implied expansion of coverage over a conventional skylight they will allow much more light to enter. Also, because walls tend to run the length of the house under the hip, a large amount of light can be cast on the central wall and reflected into the room. This kind of bright, even light will support a broad array of house plants.

Window walls

Bay or bow windows

Extended windows

Skylights

Clerestory windows

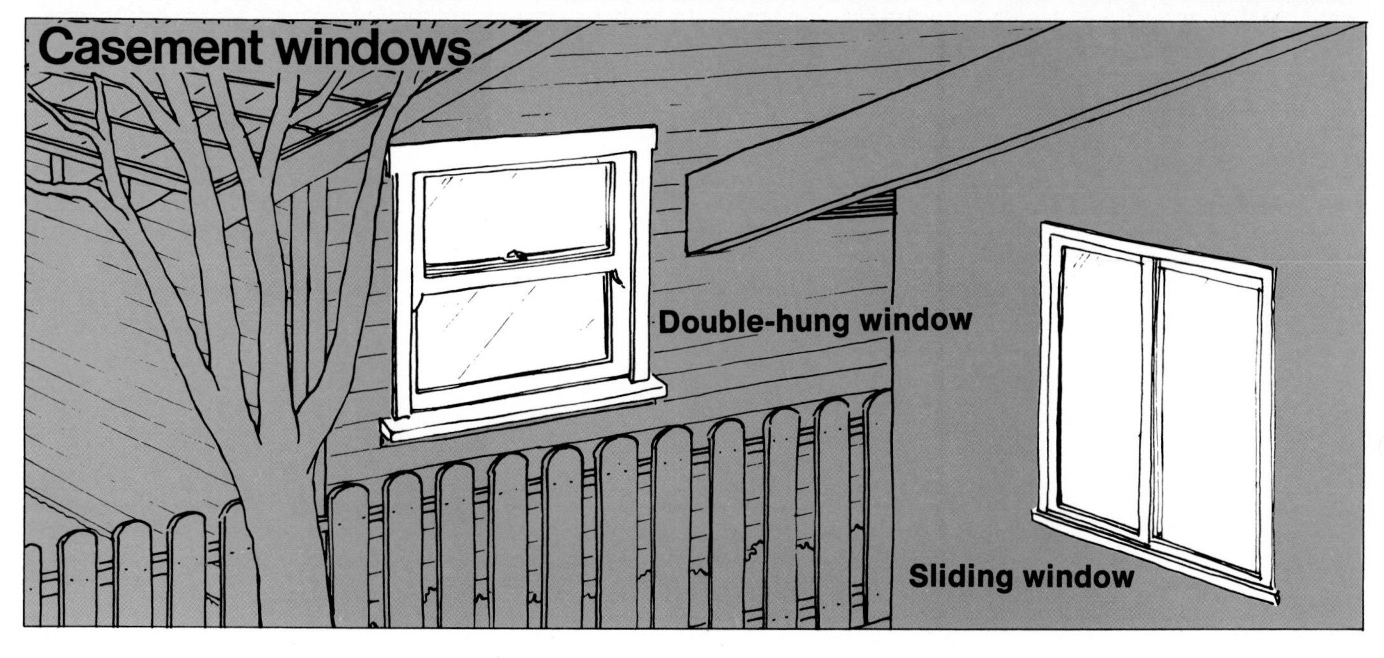
Casement windows
Double-hung window
Sliding window

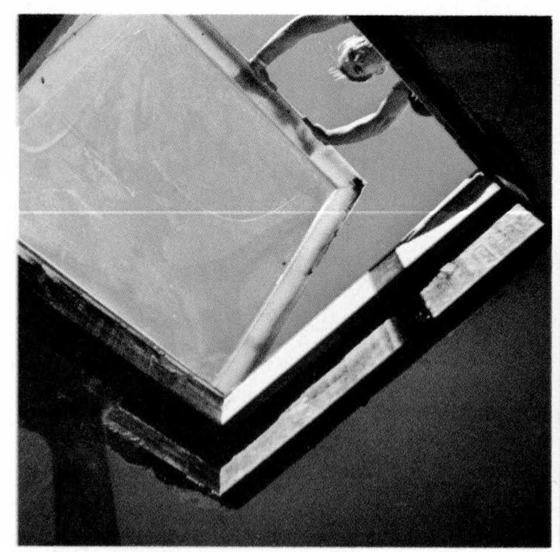
Position the bubble onto curbs on roof.

Screw the bubble to the curbs.

Protect room by hanging plastic sheets.

How to install a skylight

Skylights are widely available in both glass and acrylic plastic. The most common forms are flat panels, pyramids, double pitch types and "bubbles." They are made in various translucencies and tints. While many installations must be custom built from field dimensions, most home installations can be made from stock sizes from a dealer's shelves. The skylight shown above is 48 inches square and costs about $130.00 retail.

There are various considerations in planning for a skylight

✓The amount of light to be admitted. This is a function of the size of the skylight, the translucency of the glass or plastic and the depth of the light well, if any.

✓Where the sun will fall when the skylight is in place. In the United States skylights will light the north sides of rooms and the sunlight will move from the west side of the room in the morning to the east side in the afternoon. Be aware that direct light will fade some furniture fabrics and rugs and that some plants can tolerate only a little sunlight, if any.

✓The interior treatment. Many homes have a "crawl space" between the ceiling and the roof so the skylight must be boxed in to form a light well.

Steps in skylight installation:

✓Determine skylight size and location.

A clerodendrum has bloomed in this skylight for several years.

Skylight in kitchen is enjoyed by the cook as well as the plants.

Build inside light well of plywood.

One removable wall for attic access.

Prime and paint completed light well.

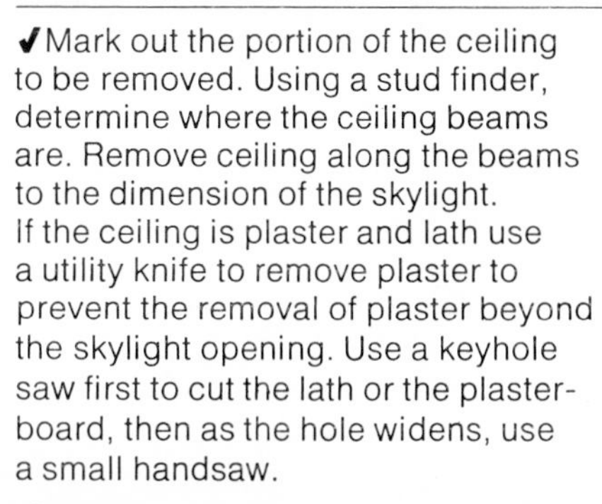

✓Mark out the portion of the ceiling to be removed. Using a stud finder, determine where the ceiling beams are. Remove ceiling along the beams to the dimension of the skylight. If the ceiling is plaster and lath use a utility knife to remove plaster to prevent the removal of plaster beyond the skylight opening. Use a keyhole saw first to cut the lath or the plasterboard, then as the hole widens, use a small handsaw.

✓Using a carpenters square or level carry the line of the corners up to the roof to mark that opening. Drive nails through to mark the four corners then remove the roof, again first using a keyhole saw and graduating to a handsaw or electric saw.

✓Next remove the beams from the opening, cutting them away *beyond* the opening size to correspond to the size of the headers that will box in the opening at both the ceiling and the roof, securing the cut ends.

✓Build a 2 x 6 curb on roof to receive the skylight, sloping it to the slope of the roof or leveling it according to taste.

✓Using roof patching compound seal around outside.

✓Place bubble on curbs and nail or screw to the curb.

✓Build light well inside of plasterboard or plywood, flush with ceiling. Trim with moulding or tape.

✓Fill, prime and paint.

Space for a hanging garden was built into the light well.

Skylight is supplemented by incandescent light on the weeping fig.

Window gardens

Prim rows of potted plants on windowsills have given way to hanging gardens, trees in tubs and masses of plants throughout the interior. However the window garden area still receives most of the light entering the indoors and should not be overlooked. Here are a few ideas for updating window gardening.

Indoor/Outdoor window boxes carry the same plant material right through the glass. Planting in the traditional outdoor redwood box is echoed in the decorative indoor box. You may choose to plant directly into the boxes, or place the pots inside the box and disguise the top with moss, bark, or stones. In either case, the indoor planter must provide a protective liner to prevent leakage.

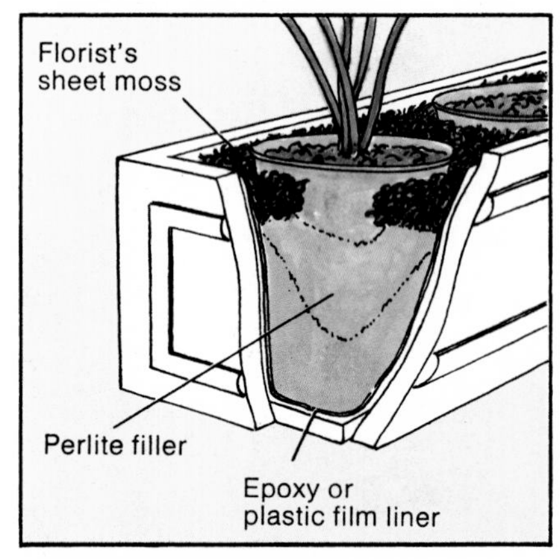

Built-in planters can be designed to correspond to the shape of the window with a custom-made metal tray to catch drainage from pots. You may add a pipe for drainage or fill the tray with pebbles and a little water for extra humidity. This idea works perfectly built over radiators.

Swinging brackets mounted on the window frame allow plants to be moved in and out of sunlight as desired. A series of brackets will allow you to create changing patterns with the plant and the plant can swing out through an open window for a breath of fresh air.

Strawberries are repeated inside and out in matching window boxes. We took a standard redwood planter, added a veneer of smooth wallboard and decorative molding. The inside was painted with waterproofing epoxy.

A built-in window planter box deep enough to hold 6 to 8-inch pots defines the area for a plant collection.

Pulleys add much flexibility to window gardens. They let you raise or lower the plant into more light or into filtered shade. As an aid to the gardener they allow plants to be lowered to levels that are convenient for watering, misting, or grooming.

Consider large pulleys connected to a long bar or square frame from which several plants can be hung in staggering positions. One simple operation lets you raise or lower an entire hanging garden at one time.

Pulleys allow Boston ferns to live up in the light and be lowered down for easy plant care.

Modifying natural light

The amount or distribution of light entering the home through a 'window' may not meet the requirements of particular house plants. However there are a number of techniques and devices that can be employed in the average household, most requiring less cost than care.

Although some of the ideas presented here are self-evident (they should be taken as reminders), they may make the literal difference between night and day in a room full of plants.

OPEN THE CURTAINS, SHADES, SHUTTERS. One of the most obvious pieces of advice is to open the curtains or shutters and raise the shades each morning on rising. Or even better, open the living and dining room curtains upon retiring so that plants may take advantage of the sun's first morning rays. Heavy curtains can bar all effective light as far as plant growth is concerned as do pulled shades. With shutters a common mistake is to adjust the movable louvers to one position, forgetting that the sun's angle changes throughout the day. It's better to open the shutters completely!

PRUNE OUTDOOR PLANTS. Large shrubs or trees may block sunlight all or part of the day. Mature lindens, acacias, some oaks and avocado trees can steal all of the light away if we let them! Thinning and pruning can admit the light that your prize Indian laurel fig needs and will improve the outdoor tree's health too. Sometimes the problem may be solved by tying branches back to let needed light pass. This can often be done in such a way, using dark-colored wire, that your rearrangement is not at all evident. You may want to consider removing a tree that has passed its prime and is in decline and replacing it with a smaller specimen.

KEEP WINDOWS AND PLANTS CLEAN. It's surprising how much light can be blocked from plants when the glass is dirty. Regular window washing is important to the plants' health. But it doesn't do much good to have sparkling clean windows if the foliage is dusty. Washing or wiping dust from leaves can add considerable light. (See page 20.)

CONSIDER AWNINGS AND OVERHANGS. When constructing or remodeling your home, remember that exterior architectural features can block the amount of light that gets inside. Select awnings that can be raised or lowered easily and can help modify light falling on plants indoors.

Interior light modification

The amount of light within a room can also be modified. This would entail redistributing the light that enters the 'window' and may be crucial in a room with a small opening. A large window naturally admits more light than a small window; there is more direct sunlight and therefore more internally-reflected light within the room. The intensity of sunlight on a windowsill plant in either window will be the same. But there is a difference in intensity as one moves away from the glass. The brightness of the light decreases much more rapidly in a small-windowed room than one with a large window. A 2,000 f.c. reading next to a glass window wall may decrease to 1,000 f.c. as one moves ten feet back into the room. That same 2,000 f.c. might be expected to decrease to as little as 200 f.c. the same distance from a common double-hung window—down to only 20% as much! There are several tactics the homeowner may employ to guard against footcandle-loss.

SHEER CURTAINS can act as thousands of tiny reflectors, dispersing moderately strong light into areas further away from the window than it would reach otherwise. While they will decrease the intensity of light in the sunlit area, they will increase the amount of light farther back into the room because of their reflective quality. A white nylon net curtain might be expected to decrease direct sunlight from 8,000 to 2,000 f.c. in the area immediately behind the window, but increase the readings from 125 f.c. to 250 f.c. five feet away.

WALL COVERINGS. One of the newer products on the market is aluminized Mylar, a thin plastic sheet to which has been applied a thin coat of aluminum by a vacuum process. This material is highly reflective and provides a good method of redistributing light in the darker parts of rooms. A related product is metalized wallpaper, which doesn't reflect as much light as the aluminized Mylar, but may be easier to accommodate to the interior design.

Light-hued paint reflects more than darker tones. White is most effective, of course, and can make the difference of 150 f.c. over a dark room with the same light source. Matte white paint will refect three-quarters or more of the light that strikes it—see the accompanying table showing the reflective quality of commonly used materials.

Mirrors and mirrorized walls can do yeoman's duty in directing and distributing light in an otherwise dark-room. Mirror tiles are readily available. In addition to making light more even over a room they will visually expand the size of your interior garden!

Evaluating available light

Now that we have a general understanding of the principles of light and exposure, and a feel for the way plants react to light, what is the next step? It's to perform an environmental inventory of our living space. This will consist of a series of check lists for each lighting situation. Refer to the charts on pages 28-41. Apply it to your existing houseplants first to make certain that each is in its optimum lighting situation. Add supplemental artificial illumination where necessary, and if possible, provide lighting in places with differing temperature ranges for the broadest possible variety of plant types. Then as you add or replace plants you will be able to make rational selections based on your actual light and heat conditions. You are finally able to cast more light on indoor plants!

Reflectance of commonly used reflecting materials

Material	Reflectance (Percent)
White Plaster	90-92
Mirrored Glass	80-90
Matte White Paint	75-90
Polished Aluminum	60-70
Porcelain Enamel	60-90
Aluminum Paint	60-70
Stainless Steel	55-65

Editor James McNair wanted more light in the bathroom of his San Francisco house. The solution was to knock out the east wall with its tiny window, extend the room three feet and add a large window that opens for fresh air circulation. Matte white tile and Mylar wall covering were added for maximum reflectance of available light. Footcandle readings tripled and plants are thriving.

Window mirror reflectors

The use of mirror surfaces to reflect direct light from the sun into a room presents many difficulties. The size of the area of light that can be reflected from a flat mirror surface can be no larger than the size of the mirror itself. Mirrors have to be directed exactly the right way, and the right way changes rapidly as the earth revolves into and away from the sunlight. Then there is the further problem of how to reflect the sun into the room without having the full glare in your eyes somewhere inside the room.

Considering these problems we were happily surprised to find one example of the use of full sun with mirror reflectors to enhance inside light. This reflector arrangement is installed on a window facing almost due south. In the summer the roof overhang excludes direct sunlight from the window. A hinged outside mirror 6 feet long and 14 inches wide along the bottom of the window reflects light upward through the window; the angle is adjusted so this light is confined to an area directly above the window. Only by bringing your face very close to the window can you see a direct reflection of the sun. Inside the room above the window a mirror of equal size reflects the light directly down and onto the plant growing area.

With these two mirrors the plant area receives over two hours of midday sun. The addition of two vertical mirrors on each side of the window to pick up and reflect earlier and later sunlight extend this to about four hours; enough to grow most outdoor garden plants.

The inside mirrors could be advantageously made a few inches wider than the outside mirrors to more fully use the shallow arc pattern of the sun's reflected movement. With three mirrors the light pattern will make three overlapping transits of the planting area. The outside mirror and the top inside mirror must be hinged so the angle can be easily adjusted. Such an adjustment to the changing sun angle should be made each week.

The outside mirror is supported by link chains with long turnbuckles at the upper wall attachment points and strong springs at the lower wall attachment points, allowing for angle adjustment. A double turnbuckle arrangement, without the springs, on each chain would provide more positive restraint against movement in strong winds.

Without the inside mirrors, just the bright patch of sunlight directed onto the white ceiling is enough to double the light intensity in the center of the room 8 feet from the window.

In winter when direct sunlight comes through the windows the combined direct or reflected light, even with the loss from the window glass and the mirrors, is the equivalent of tropical desert light.

First you have to have a window facing due south. Adding only an hour of direct reflected sunlight will double the radiant energy on plants receiving only reflected light from the sky. Maybe there are other window orientations where modifications of this method would work. Just attach a shaving mirror to a board and run it out the window—adjust it to put the bright circle of reflection on the ceiling and see where it moves for an hour or two. The least you may come up with is an unusual seasonal sundial on the ceiling.

A combination of mirrors—indoor and out—reflect light on indoor plants.

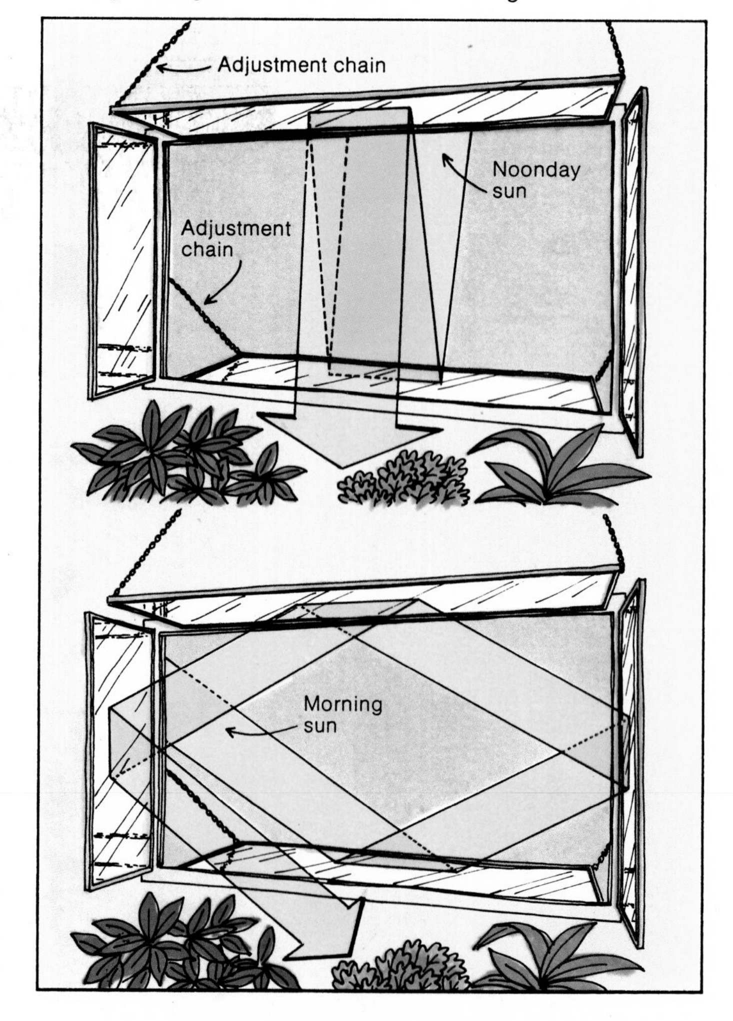

Above: We built a folding screen, painted the frame matte white and inset Mylarized panels. It did a good job reflecting light from windows across the room onto plants on table. Also it gave added dimensions to the plants. A floor-standing screen is another alternative for large plants. Below: This collection of cactus in San Francisco gains necessary additional light reflected from south and west windows onto a vertical mirrored screen.

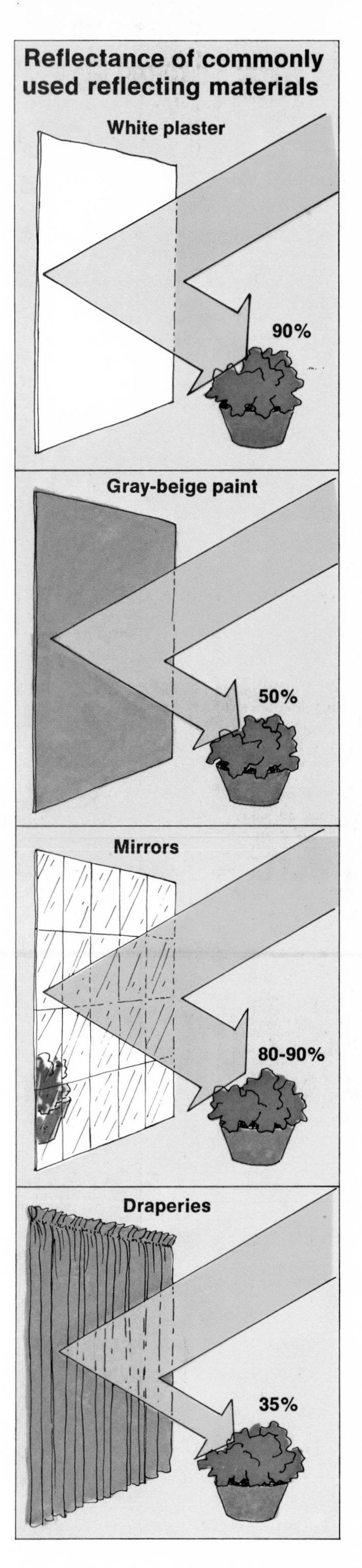

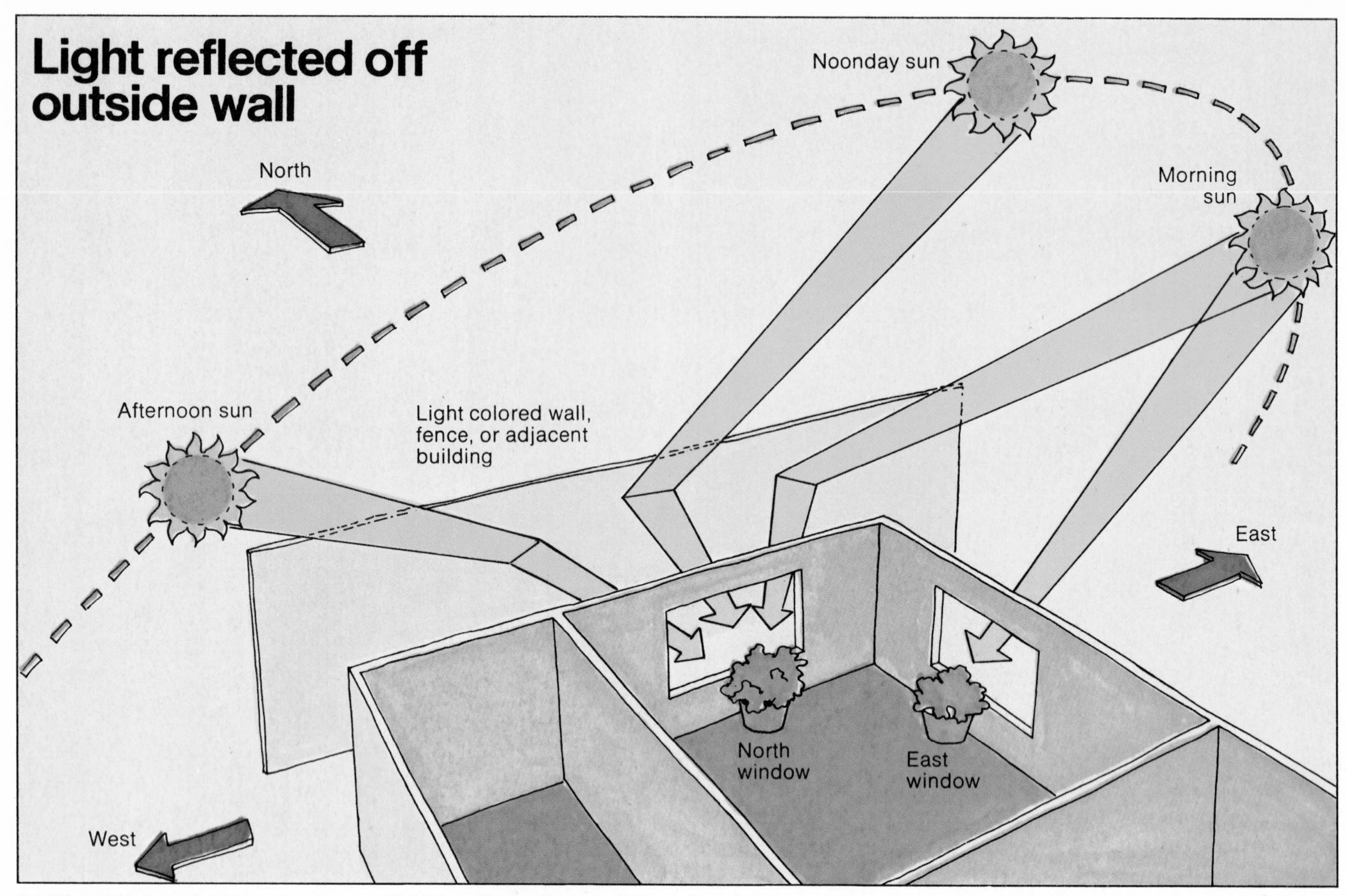

Additional indoor light is gained as sunlight is reflected from a light-colored wall or other outside reflecting surface. Total light bounced off the wall through the North window can equal the morning direct sun from the East window.

Shutters can adjust amount and direct light through windows. For maximum light, remember to open the shutters each day.

Sheer curtains serve two purposes: they filter out harsh afternoon sun and reflect a more even light through the room.

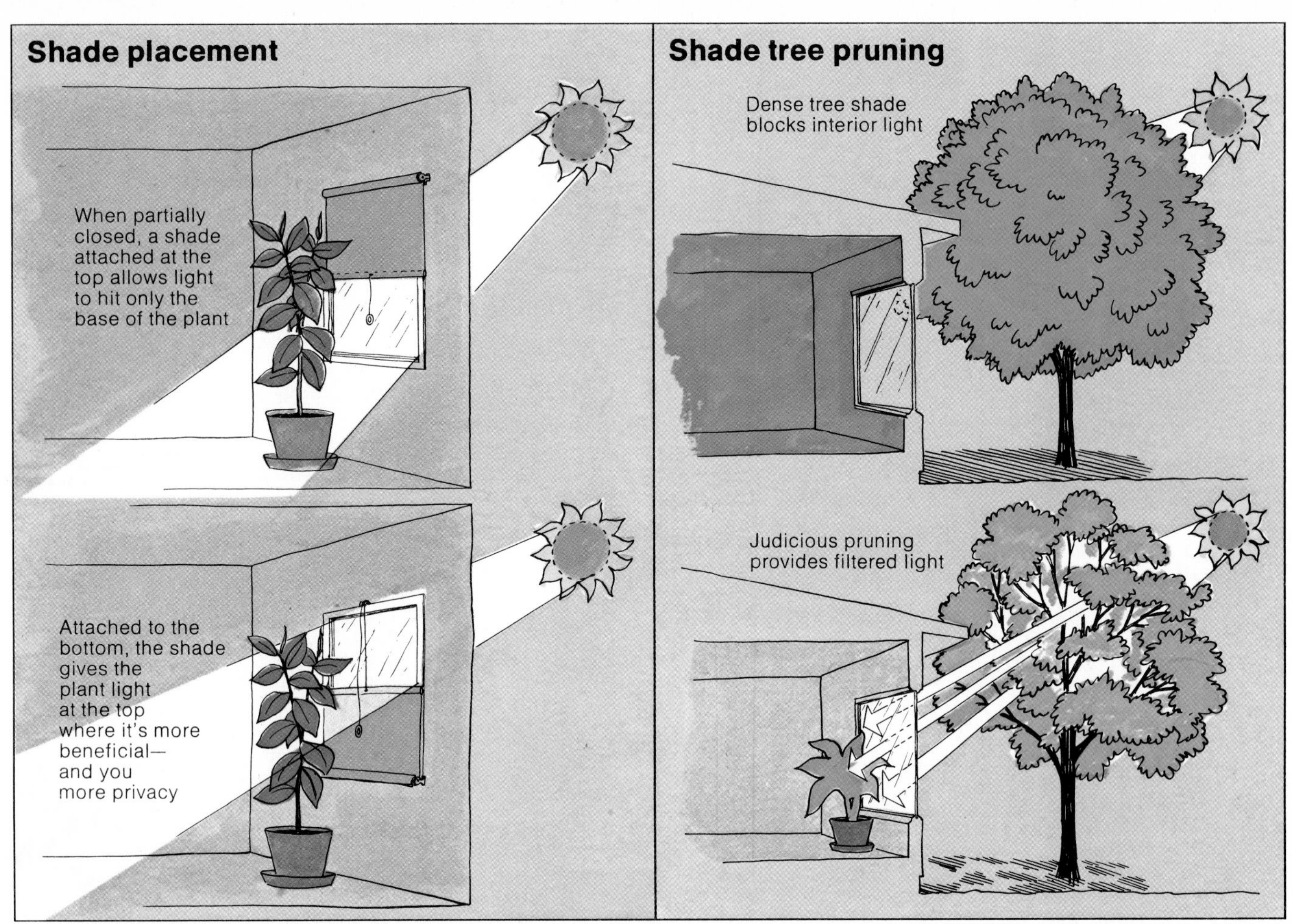
Shade placement
When partially closed, a shade attached at the top allows light to hit only the base of the plant
Attached to the bottom, the shade gives the plant light at the top where it's more beneficial— and you more privacy
Shade tree pruning
Dense tree shade blocks interior light
Judicious pruning provides filtered light

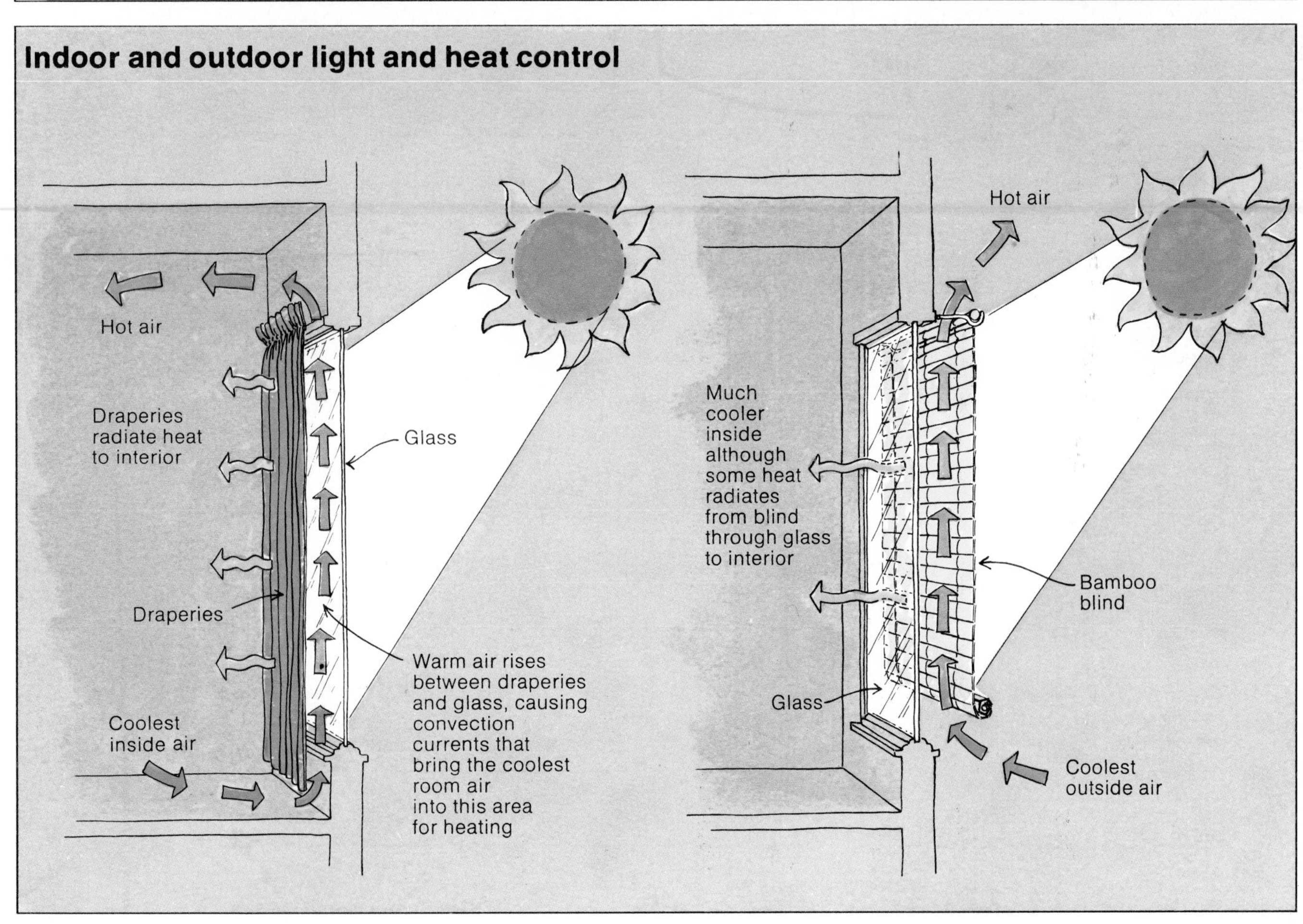
Indoor and outdoor light and heat control
Hot air
Draperies radiate heat to interior
Glass
Draperies
Warm air rises between draperies and glass, causing convection currents that bring the coolest room air into this area for heating
Coolest inside air
Hot air
Much cooler inside although some heat radiates from blind through glass to interior
Bamboo blind
Glass
Coolest outside air

You can live in a glass house

Greenhouses are traditionally divorced from the house—they are found out in the garden where plant activities often take place. But since we have moved the garden indoors, why not also move the greenhouse inside too!

The advantages of "indoor" greenhouses are obvious—not only are they ideal for growing tropical plants, they also serve as display areas where seasonal color may be proudly shown without walking out to the south forty! However, there are some problems with integrated greenhouses. Primarily, these are humidity, neatness and water.

In order to maintain proper humidity the greenhouse must be enclosed and physically separated from the rest of the house to prevent damage to furnishings. In order to do this and at the same time maintain the "tie", glass sliding doors or window walls are best employed to divide the "greenhouse" from the other interior.

Neatness is a problem. All maintenance-related activities should be hidden if the greenhouse is to integrate well with the adjacent room. Even more important are the benches and other greenhouse furniture. They should be designed in such a manner and with the appropriate materials that they are harmonious with the interior decor.

Watering or spraying should take place so that its influence isn't felt in the indoor room. Windows and doors should be firmly sealed and steps taken so that water is not tracked into the house.

Careful planning will overcome the obstacles and the added value to your home and lifestyle is really worth the effort.

Mr. Bob Locatell of Atlanta, Georgia, has done just that. He has built an 8′ x 19′ solarium next to his living room/dining room/study complex. He "leaves the sliding doors open all winter long so that, in effect, this becomes an extension of the living room." The solarium is constructed of a frame with 4′ x 8′ removable panels of 1/8″ thick clear plexiglass. This material is lighter weight than 'Thermopane' or other plate glass and each panel can be easily handled by one person.

Mr. Locatell says that "the sloped roof is a single sheet of 'Kalwall' 1 3/4″ thick, which is a plastic material that has been on the market for a number of years. It cuts excessive glare but transmits 60-70% of the light that falls on it. It has an insulating value equivalent to that of an insulated stud wall, which conserves heat." Although heat is conserved in winter by the low angle of the sun, he uses a 1200 watt electric heater with a fan, set at 60°, to prevent night temperatures from dropping too low. The panels are placed in position in late fall and removed in late spring making the area an open patio for warm weather use. He uses an indoor-outdoor carpet in winter which is rolled up and stored in summer.

Pest-control is handled by spraying Malathion in early spring and again, at the start of fall. The sliding doors are then closed for two days. Even with a large number of plants, watering takes only 30-60 minutes daily.

Bob Locatell.

Added to the home of architect Dale Durfee, AIA, is a two-story "lean-to" greenhouse room. Mr. Durfee enjoys growing primarily tropical plants. This is the view from his upstairs living quarters.

Mrs. Davis Love, of Marietta, Georgia, has a solarium for the primary purpose of adding extra living and entertaining space. "We live in it the year round and use it every day. We eat breakfast there, and most of the time, dinner. In fact, the changing seasons have their impact on the mood of the solarium." The walls, which are windows and sliding glass doors, open up half way to convert to a screened porch in the summer time. At Christmas Mrs. Love changes the green and yellow decor to all green, reflecting the season.

Maintenance isn't much of a problem. Her floor is of Armstrong 'Solarium,' which takes water and high humidity. Pots catch the water so there is no dripping. In one corner there is a 3′ x 3′ area, under a staghorn fern, that is filled with pebbles and drains to the outside. This area is used for those plants which require deep soaking. This, in turn, helps create high humidity. Orchids and bromeliads are kept nearby to take advantage of a very suitable corner microclimate.

In winter, heat is provided by five electrically heated baseboards (she thinks two would have been enough). These are set down to 60° at night, which Mrs. Love believes is the secret of her plants doing so well. "My geraniums bloom all winter when it is 40°-50° outside. The sun on the windows gets the daytime temperature up to 80°."

Mrs. Love has grown tomatoes and cucumbers in the solarium, getting a jump on spring. "By the time I can put plants outside, they're almost ready to eat." She has successfully started plants from seed but plans on building a separate "working greenhouse" to move "the clutter" away from the living space.

Mr. Richard Marell, San Francisco attorney, has built a greenhouse opening off his deck adjacent to the living room. It is constructed of redwood members with flat fiberglass panels designed to eliminate between 70%-85% of the ultraviolet rays. It houses his prized orchid collection, which creates an everchanging panorama from the living room.

Traditional in design, it has slatted wood benches which provide the air circulation so necessary for healthy orchids. His equipment consists of two small whisper fans for constant air movement, running 24 hours a day, one vaporizer, and one thermostatically controlled exhaust system, set at 78°.

Mrs. Davis Love.

Richard Marell.

Jim Gibbs.

Mr. and Mrs. W. E. Freeborn.

Mr. Jim Gibbs, Atlanta, Georgia, has built a 'Thermopane'-paneled solarium off of his dining room. Designed for low maintenance, it has a slate floor which is ideal in a wet greenhouse situation. He waters less than one hour a day, however, since not all pots need watering and the ground beds retain moisture.

The solarium is used as a family eating area and for entertaining. It is enormously popular with visitors who are "automatically" drawn to this area. Many friends have had similar structures built since seeing the solarium.

Mr. and Mrs. W. E. Freeborn of Atlanta, Georgia, enclosed the small back porch area off their kitchen door. It has a corrugated fiberglass roof and the side is adorned with a stained glass window. They call it the "green room" and here Mr. Freeborn comments about it:

"In the first place the "green room" does not have any heat in it but gets the heat from the house and the door stays open practically the year round. We have found a few days in some winters when it was necessary to have additional heat but there is electricity in the "green room" and we use an electric heater for this purpose.

"The bench is warmed by a cable.

"It gets used mostly for green plants although we do have some blooming things there and the things that have done best have been Thanksgiving Cactus, Begonias, and Bougainvillea.

"The grow and bloom lights on the table give excellent results and are used for Violets and a few annuals. We also think that this helps both cuttings and seedlings.

"Of course, that area gets a great deal of sunlight.

"My observation of Mrs. Freeborn is that the things that she has enjoyed the most are (1) growing seedlings, (2) growing plants from cuttings, (3) caring for friend's plants which aren't doing well."

Mr. and Mrs. Freeborn's "greenroom" proves that "glasshouses" need not be large or elaborate to be effective. There is a whole range of ready-mades on the market from small window units to large commercial-sized fully-equipped greenhouses. In some cases there may be nothing available that suits the needs of the indoor gardener. If so, an architect can provide a custom solution. On page 72 are some of the types that should be considered.

Types of greenhouses

WINDOW EXTENSION. This unit fits into the window, utilizes indoor heating and provides extended growing space.

LEAN-TO. The lean-to creates a walk-in nook. It allows the introduction of larger plants than the window box above and the use of benches and some maintenance equipment, if desired.

GREENHOUSE EXTENSION. This type is a complete greenhouse appended to the house. There are usually doors or vents for control of moisture.

SUN ROOM. Like the greenhouse extension, the sun room is a separate room, but it is more like an "outdoor" living room with furniture. It normally has glass walls and a ceiling with shades that can be pulled to control direct light. Often these rooms have tile floors to minimize the watering and maintenance problem.

DECK GREENHOUSE. For city-dwellers, deck greenhouses may be the only gardening possibility. Often, otherwise unusable decks can be transformed into an exotic jungle right off the living room. For renters it is possible to build a mobile greenhouse that can be easily taken apart and reassembled elsewhere. See illustration.

Prefabricated units or components for building your own design may be secured from these greenhouse suppliers:

Aladdin Industries
P.O. Box 10666
Nashville, TN 37210

Aluminum Greenhouses, Inc.
14615 Lorain Avenue
Cleveland, OH 44111

Casa-planta
16135 Runnymede
Van Nuys, CA 91406

Feather Hill Industries
Box 41 Zenda, WI 53195

Greenhouse Specialties Company
9849 Kimker Lane
St. Louis, MO 63127

Hansen Weather-Port
300 South 14th
Gunnison, CO 81230

Ickes-Braun Glasshouses
P.O. Box 147
Deerfield, IL 60015

Lord and Burnham
Irvington, NY 10533

National Greenhouse Company
P.O. Box 100
Pana, IL 62557

J. A. Nearing Company
10788 Tucker Street
Beltsville, MD 20705

Sturdi-Built Manufacturing Co.
11304 S.W. Boones Ferry Road
Portland, OR 97219

Texas Greenhouse Company
2717 St. Louis Avenue
Fort Worth, TX 76110

Verandel Company
Box 1568
Worcester, MA 01601

Above: Avid orchid collectors Harold Ripley and Bob Hoffman added two greenhouses to hold their 3000 plants—one for cool-loving species off the dining room, left, and a warm house off the kitchen. They installed a window unit, right, in the bathroom for additional plants. Below left: Mrs. Gerry Hull converted a long hallway into a sunroom by adding skylights. Below right: The Gene Baumgarten's outdoor patio received strong winds from the San Francisco bay, making it impossible to use. Designer David Adams enclosed the patio, creating a glasshouse where the Baumgartens spend much of their time.

Deck (or balcony) greenhouse (see drawing on page 72)

1. **To avoid a possible hassle,** get your landlord's permission and check local building code regulations to see if this kind of a structure is permitted on a balcony or patio.

2. **Planning.** If your balcony or patio is 4½′ x 8′ or larger and your outside door frame 77″ wide or smaller you can use the dimensions shown here. If not, or if you prefer a larger or smaller greenhouse, you will have to modify the plan and material sizes.

3. **Basic construction.** Our green house consists of ten prefabricated frame panels, covered with "plastic," and screwed together on a free-standing wood foundation. Construction of all panels is the same, only the sizes change and the roof panels have an extra 2″ x 2″ on one end for added length. The door panel is optional, of course, but nice to have if your balcony is larger than the greenhouse. If you opt for no door you'll have to provide for easy removal of the covering on the lower half of one panel so you can crawl in and out during assembly and whenever else you need to get outside the greenhouse.

Materials Needed

Wall & Roof Panels

14 pcs. 1″ x 2″ x 72″
16 pcs. 1″ x 2″ x 22½″
4 pcs. 1″ x 2″ x 34½″
6 pcs. 1″ x 2″ x 48″
8 pcs. 2″ x 2″ x 22½″*
2 pcs. 2″ x 2″ x 34½″*
2 pcs. 2″ x 2″ x 24″
1 pc. 2″ x 2″ x 36″
2 pcs. 2″ x 2″ x 24″
1 pc. 2″ x 2″ x 36″
1 pc. 1″ x 1″ x 87″

Foundation

2 pcs. 2″ x 4″ x 87″
2 pcs. 2″ x 4″ x 80½″
2 pcs. 2″ x 4″ x 46″

Door

2 pcs. 1″ x 3″ x 70″
3 pcs. 1″ x 3″ x 17″
4 pcs. ¼″ x 6″ x 6″ plywood triangle corner braces

*These pieces may be 1″ x 2″ if you are not using rigid plastic covering material.

Hardware

1 pc. 1″ x 1″ x 87″ Aluminum "angle-iron"
4 doz. 1½″ x #8 wood screws
9 – 3″ x #12 wood screws
6 – 2″ x 2″ metal "L" brackets and screws
2 doz. 3″ metal mending plates and screws
2 pcs. 8′ long vinyl baseboard
40 3″ x 3″ metal "L" mending plates with screws (optional)
52 ¾″ x #6 wood screws
2 lbs. galvanized 6d box nails
1 pr. 3″ butt hinges with screws (for door)
1 screen door hook to latch door

Panels

Assemble ten wall and roof panels like this.

Pre-fab Units to Build

5 – 24″ x 72″ wall panels
1 – 24″ x 72″ door panel
1 – 36″ x 72″ wall panel
2 – 24″ x 48″ roof panels
1 – 36″ x 48″ roof panel
1 – 84″ x 51″ foundation frame

Door

Assemble door of 1″ x 3″s laid flat with ¼″ x 6″ x 6″ plywood corner braces. Use ¾″ #6 wood screws to hold corner braces. Fit door into panel and attach hinges and hook.

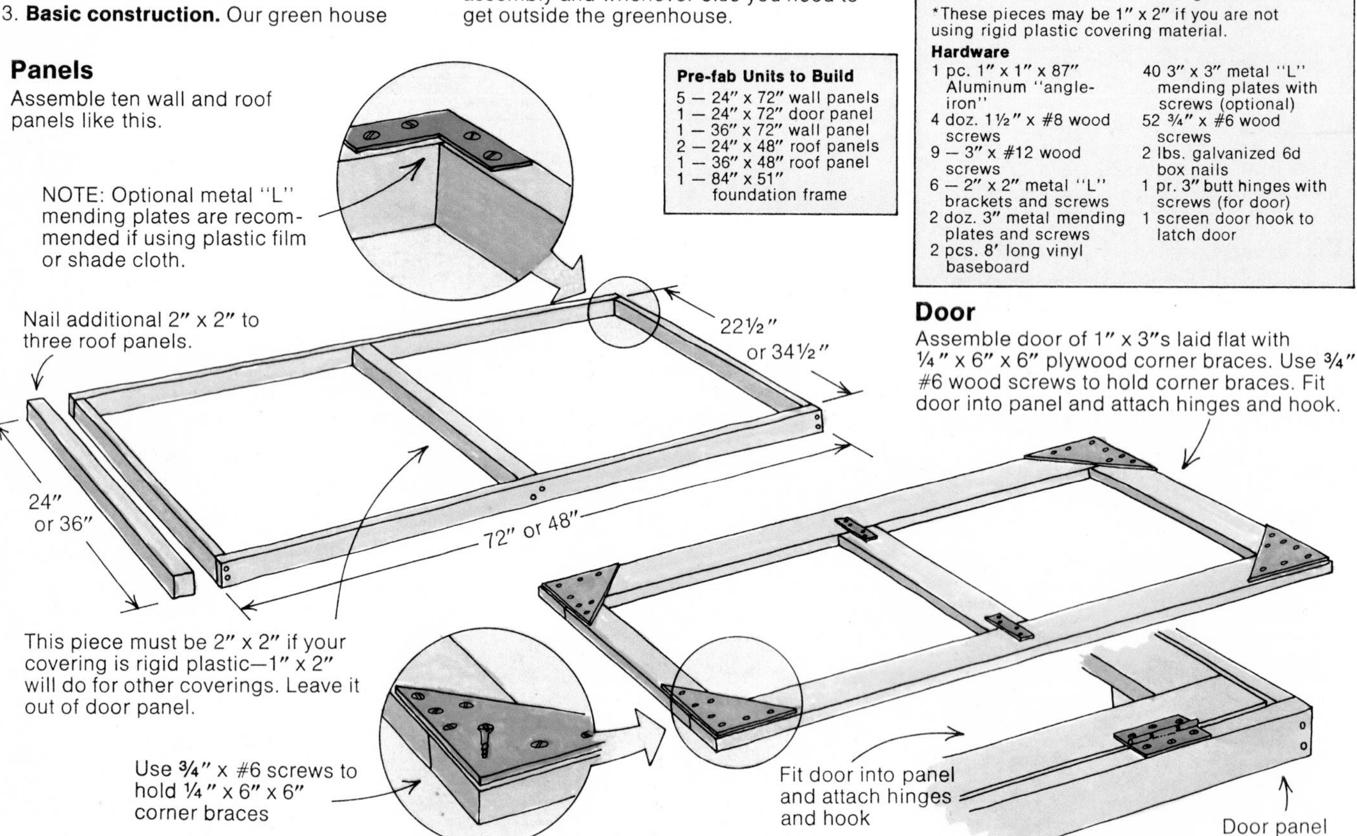

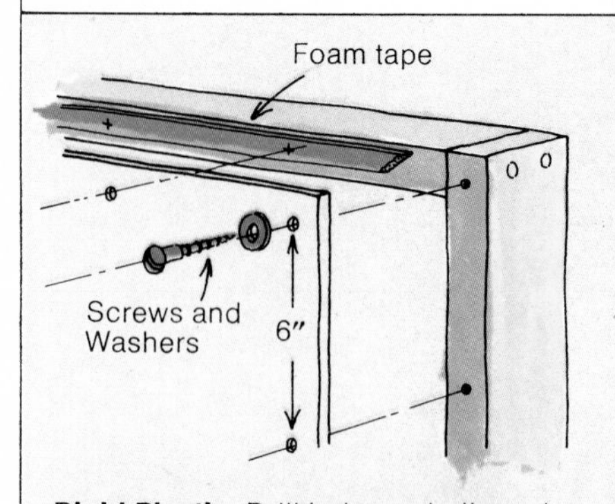

Rigid Plastic. Drill holes as indicated and screw the plastic sheets gently, not too tight, to each panel. Foam tape or a bead of putty or caulking compound will seal the edges and allow for some expansion and contraction of the plastic.

Materials needed

Plastic
2 pcs. 36″ x 36″ (sides)
10 pcs. 24″ x 36″ (sides) (12 if no door)
1 pc. 36″ x 48″ (roof)
2 pcs. 24″ x 48″ (roof)
2 pcs. 34″ x 34″ (door)
1 pc. 15″ x 48″ (see 6)
Insulation tape, caulking compound, or putty
2 gross galvanized ¾″ #6 round head screws and washers
Aluminum screen, 1 pc. 15″ x 50″

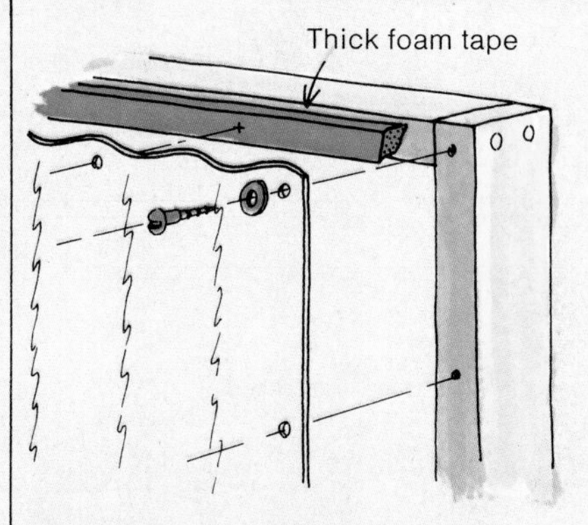

Corrugated Fiberglas. Drill holes as indicated and screw the plastic to each panel with the foam tape in position—especially at the top and bottom to fill in the spaces between corrugations.

Materials needed

Fiberglas
1 pc. 36″ x 72″ (side)
5 pcs. 24″ x 72″ (sides) (6 if no door)
1 pc. 36″ x 48″ (roof)
2 pcs. 24″ x 48″ (roof)
1 pc. 34″ x 70″ (door)
1 pc. 15″ x 48″ (see 6)
Foam tape
2 gross galvanized ¾″ #6 round head screws and washers
Aluminum screen, 1 pc. 15″ x 50″

Start at centers of sides and ends . . .

Plastic film or shade cloth. Cut sheets of material somewhat oversize. Lay the panel frame on the floor and lay the covering over it. Put a tack or staple in the center of one end, pull tight toward the center of the other end and tack it. Tack the centers of the sides the same way, then alternate stretching and tacking each end and side toward the corners as you move around the panel. When it's all tacked trim off excess material. This method gives you a tight, wrinkle-free covering.

Materials needed

200 sq. ft. of film or cloth (about 60 lineal feet of 40″ width)
Staples and staple gun or tacks

4. **Covering.** You have several choices for covering your greenhouse depending on your climate, the appearance and durability you want, and the amount of money you want to spend. Rigid, clear plastic storm window material or plexiglas is best for the colder climates. It's durable and transparent like glass and also the most expensive covering material here.

Filon or corrugated fiberglas is also good for cold climates and is very durable, but it too, is expensive and is only translucent so you can't see out and your neighbors are unable to see into your beautiful garden. It also comes in a choice of colors.

Plastic film is pretty transparent and inexpensive, but tears easily and may last only one season, especially where wind is a factor.

In extremely cold climates you may want to double-glaze your greenhouse. Adding plastic to both sides of each panel would add greatly to the insulation.

In sub-tropical climates you may want to cover your greenhouse with nursery shade cloth (Saran cloth) for shade, air circulation and insect protection.

5. **Assembly.** When the panels are finished it's time to put it all together.

The foundation of 2 x 4's is next—build it so it looks like the drawing. You may need a temporary brace (as shown) until you get the first panels in place. Use a level and/or framing square to be sure the joints are all square so the panels will fit snugly.

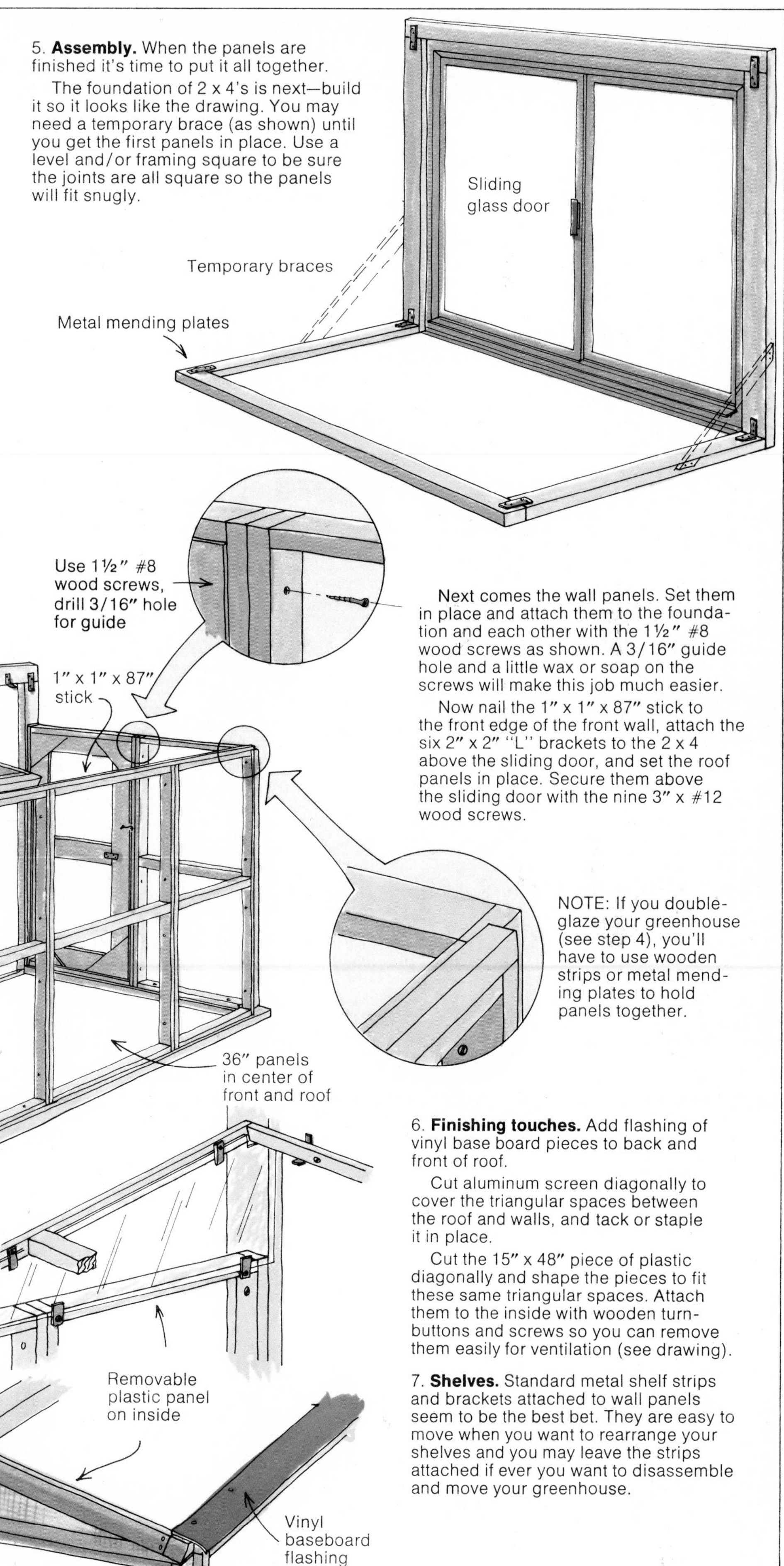

Next comes the wall panels. Set them in place and attach them to the foundation and each other with the 1½″ #8 wood screws as shown. A 3/16″ guide hole and a little wax or soap on the screws will make this job much easier.

Now nail the 1″ x 1″ x 87″ stick to the front edge of the front wall, attach the six 2″ x 2″ "L" brackets to the 2 x 4 above the sliding door, and set the roof panels in place. Secure them above the sliding door with the nine 3″ x #12 wood screws.

NOTE: If you double-glaze your greenhouse (see step 4), you'll have to use wooden strips or metal mending plates to hold panels together.

6. **Finishing touches.** Add flashing of vinyl base board pieces to back and front of roof.

Cut aluminum screen diagonally to cover the triangular spaces between the roof and walls, and tack or staple it in place.

Cut the 15″ x 48″ piece of plastic diagonally and shape the pieces to fit these same triangular spaces. Attach them to the inside with wooden turn-buttons and screws so you can remove them easily for ventilation (see drawing).

7. **Shelves.** Standard metal shelf strips and brackets attached to wall panels seem to be the best bet. They are easy to move when you want to rearrange your shelves and you may leave the strips attached if ever you want to disassemble and move your greenhouse.

Indoor seasons

Shift plants with the sun and enjoy year-round blooms, vegetables or foliage. Ways to make moving a little easier.

The changing seasons have an immediate effect indoors. The amount of light entering rooms through 'windows' increases after the winter equinox and decreases after the summer equinox. The angle of direct sunlight changes as the sun moves north in the spring and south in the fall. Since the winter sun is lower in the sky, plants that may have received only indirect light in the summer may be sunlit in the winter.

As lighting conditions change plants will become 'out of focus.' To make up for this change in light *shift indoor plants as the seasons change.*

Plants that grow well in an east or south summer window may be best moved into a western exposure in the winter. Plants that are marginal in times of brightest light may require the aid of supplementary artificial illumination in order to maintain robust health.

The best way to get to know exactly what your light conditions are is to note the extent of direct sun penetration into a room and measure the foot candles (see page 26). Each room should be canvassed in this manner—you'll be surprised how light some rooms are and how dim others can be! This process should be repeated at two-month intervals and more often where light is marginal. The more northern latitudes will register the most change between summer and winter. Your inventory, judiciously followed, will keep you in step with the changing seasons.

Plant mobility

Like wine, some plants 'travel' better than others. Weeping fig *(Ficus benjamina exotica)* and gardenias, for example are very sensitive to changed position—they can easily lose all of their leaves after being moved. It takes a while to recuperate from the shock of moving before 'coming back' with new leaves.

When bringing a plant into your home from the nursery be particularly sensitive to its response. If it wilts, this may be due to a drastic change in the environment from a very humid condition, or from a vastly different light, etc. Water it unless the potting mix feels moist. (Overwatering can produce symptoms identical to dehydration wilt.) Even more important at this time is to increase humidity to reduce further loss of moisture. (See page 19.) Be sure to check the foliage (and flowers if any) to determine that the wilt is not caused by infestation of some parasite, such as aphids, mites, white fly, etc. Start it off in a location with the brightest indirect light possible. *(Never put a plant in direct sunlight when it is wilted.)* Some plants, such as coleus, piggyback *(Tolmeia menziesii)* and campanula will droop the moment they feel too thirsty. Try to avoid this wilting from lack of water because it reduces a plant's vitality.

Keep plants away from cold drafts or hot air currents. Foliage should be misted often. Keep soil evenly moist and do not feed for first month. After a few weeks it can be moved to areas with proper light for that species' requirement. Symptoms of poor light do not usually become evident immediately; but some plants such as *Asparagus sprengeri* and *A. sarmentosus* will register their complaint about inadequate light within a week to ten days. Their fern-like foliage begins to yellow and then to fall. Swedish ivy *(Plectranthus australis)* with low light becomes leggy. Adjust misting and watering to the plant's needs and begin a regular program of feeding. (See pages 19, 20.) With time a plant will certainly let you know whether or not it is happy!

Mobilize your indoor garden!

In creating the most dramatic effect it takes less time to move large specimens than a host of small pot plants. Your Environmental Inventory may lead logically to a plant redecorating scheme—because you may have to move some plants, why not go all the way and develop a seasonal station for each plant? Check the example on page 80.

Make it easy to move plants as the seasons and lighting conditions change. Here are some ways:

Pots are the most obvious technique designed for easy movement. They are light weight and come in sizes small enough to put almost anywhere—on tables, sideboards, or in bookcases. See page 18.

Wooden platforms on casters with a detachable tote rope provide an easy way to move large plants that can't be picked up. These can be easily moved as the season demands. They also keep terracotta saucers off the floor. An alternative is to mount casters directly onto decorative wooden boxes that can be moved anywhere easily.

Plant perambulators or carts loaded with plants can make a permanent display in the living room—permanent, that is, until Old Sol changes the

This kitchen garden has adjustable shelves that aid in the seasonal shift of plants to the sun.

seasons and it's time to perambulate into the dining room to recapture afternoon sun. Carts can be easily constructed so that they blend in with even the most traditional decor—fine wood veneers or chrome and glass can make a plant cart welcome in any living room.

Pulleys allow great mobility—vertically! They make it possible to move plants up into the sun's direct rays in a sunny window in summer and back down in winter. In a skylight, the reverse—plants that would fry in the collected heat of a summer skylight can be lowered into the cool of the room via pulley. And pulleys allow changing the composition if there are several plant pulleys in a group. With pulleys your ivy, heart-leaved philodendron or wandering Jew can be raised as they grow toward the floor.

Brackets. Plants mounted on swinging brackets can be swung into or out of direct sunlight as the day or the season progresses. Cast iron brackets in the traditional mode or often more popular varieties can be mounted on a window frame for easy environmental change. Plants can even be swung out of an open window to get a breath of fresh air!

The winter garden

A sunny southern kitchen window can be the setting for a mini-vegetable garden. Lettuce, like the 'Atom' tomatoes and radishes can be produced in great abundance. Try other vegetables for the pleasure of watching how they grow and enjoying their beauty, but don't count on great indoor harvests.

The garden might occupy a countertop, or even be multi-leveled if the windows are large enough. Do not despair if you lack counter space. An herb garden can be grown on narrow shelves set into the window.

You can get a jump on spring by starting seedlings, cuttings, annuals and herbs indoors before the outdoor weather permits. Vegetables and flowers can also be started under lights for transplanting outdoors when the weather is warm enough. Because seedlings grow so quickly incandescents can be used in the early stages of this process.

Flowers can be grown in a basement or closet light garden and moved upstairs when in bloom for great effect during winter. Enjoy pots of petunias and alyssum at Christmas time. While low-growing varieties are the easiest to produce, taller plants can be developed by moving the fluorescent lamps (See page 90) up as the plants grow.

If you want to concentrate your garden and enjoy spring-flowering bulbs, even in the bleak winter months, consider the florists' method of forcing bulbs. The steps involved are quite simple:

1) Choose a container—clay pot, ceramic bowl or what-have-you that is at least twice as high as the bulb. (This will allow for adequate root development.) The container must have a drainage hole at the bottom. Use a light planter mix. Plant the bulbs shoulder to shoulder for full effect. The tips of the bulbs should protrude above the soil surface. Water thoroughly by setting container in a pail of water and letting it soak until the surface of the soil feels moist.

2) Place containers where they can get 10 to 12 weeks of 'cold' treatment-temperatures between 40-50 degrees. Any spot that's cold and dark is satisfactory. An unheated cellar or vegetable storage unit is ideal. A covered cold frame outdoors will do the job.

Outdoors pots should be covered with peat moss or sawdust or shredded polystyrene. This material never freezes, is light weight, allows water to pass through rapidly and the bulbs can be inspected at any time during the storage period. The purpose of the storage period is to give the bulbs the chilling they require to develop a strong root system.

Roots require moisture for growth. Soil should be moist when containers go into storage and kept moist throughout the storage treatment.

3) When the sprouts of the bulbs are 2 to 5 inches high and the roots can be seen at the drainage hole, place the container in a cool 60-degree room. After a week or two they are ready to take normal room temperatures.

Plant bulbs anytime from October to December. It's a good idea to stagger your planting to get a continuous supply of flowers. Plant October 1 for late January flowering; in mid-October for February color; in mid-November for March and April flowering.

Care must be taken in transferring plants grown indoors to their permanent outdoor growing place. They should be 'hardened-off' by being moved to a sheltered outdoor spot away from direct sun and wind and watered less frequently. They can be transferred to their desired location within seven to ten days.

The year-round garden

'Stage-manage' a year-round plant show. Provide a solid support cast of foliage plants to act as a backdrop—a foil for the show-offs. Bring 'outdoor' plants indoors for short periods—a handtruck can assist in moving large trees in 2' boxes to create a dramatic effect for parties. Bonsai and other small plants can be kept in decorative containers for dinner party table centerpieces. Rhododendrons and other outdoor flowering plants can be grouped indoors for several days without harm. Potted annuals and perennials can be grown on the 'south forty' and brought indoors at the height of their bloom to beautify a dining room. These are visiting star performers and will add seasonal color and interest to your indoor garden.

Sally Hildebrand (opposite page) harvested cherry tomatoes in time for Easter. This all began in late winter when transplants from the artificial light garden were potted into hanging containers (top photo above). Plastic tubing was planted in the middle as a reservoir for water. Young plants grew upward (lower photo) before cascading into trailing mature plants.

Plant movers

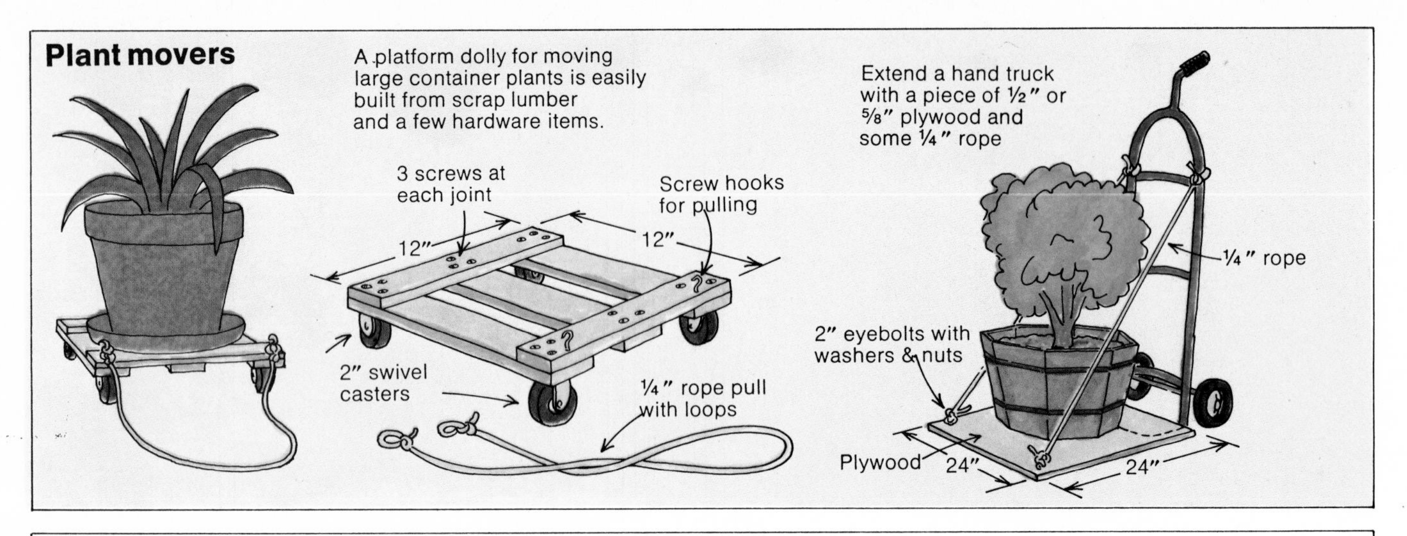

Redecorating with plants by season

Move your plants as well as furnishings according to the season. Here is a hypothetical living room (shown without doors and with windows facing all four directions) illustrating what may be done with each exposure in each season.

Key: A. Sofa & chairs **B.** Coffee table **C.** Fireplace hearth & mantle **D.** Table **E.** Christmas tree

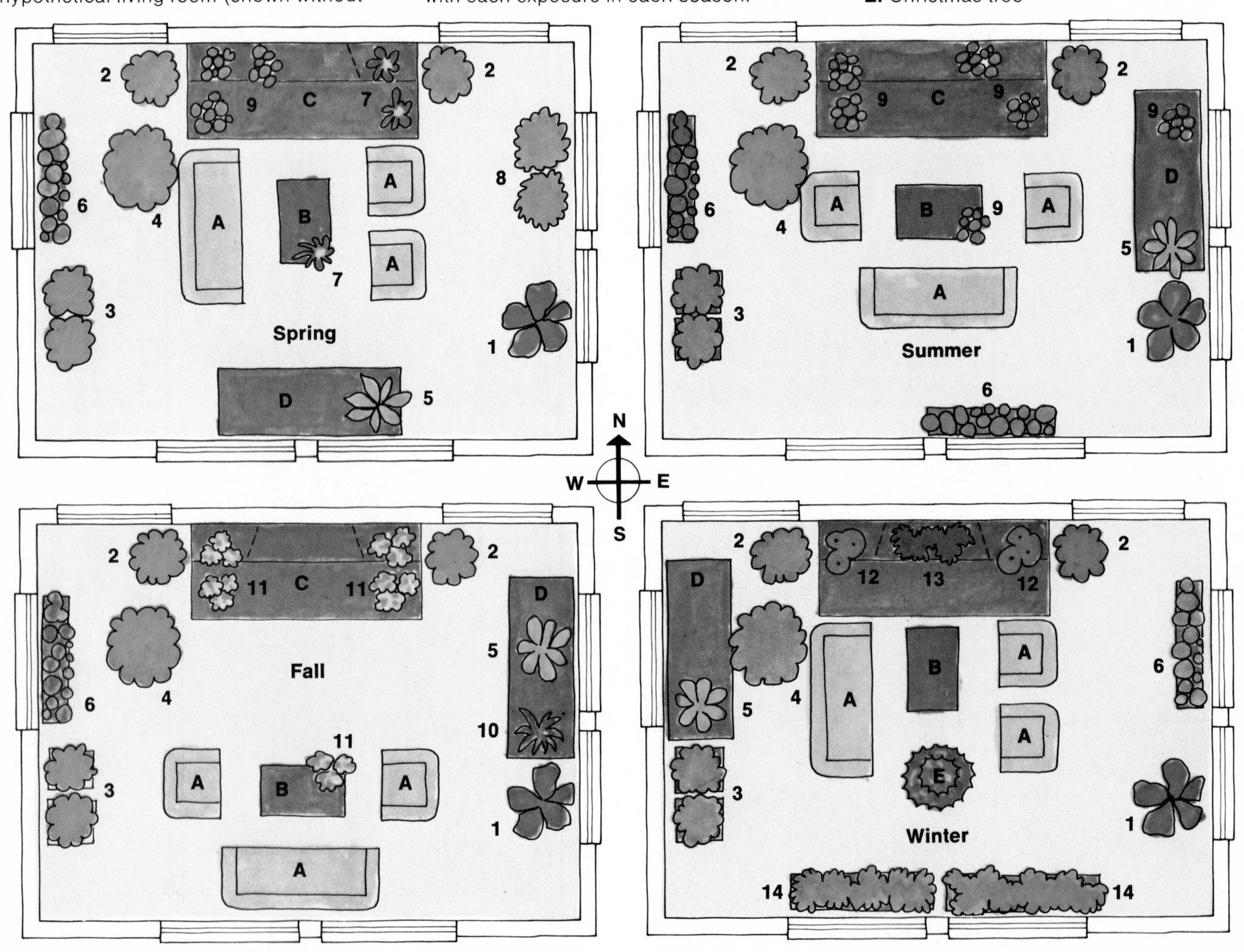

PLANTS: **1.** Fiddle leaf fig (Ficus lyrata takes direct, filtered sun from the east all year. **2.** Cast iron plant (Aspidistra elatior) takes low light of northern exposure. **3.** Hawaiian ti (Cordyline terminalis) thrives in southwestern exposure with bright, not direct sun. **4.** Weeping fig (Ficus benjamina) needs bright, indirect light; might drop leaves if moved. **5.** Mother-in-law's tongue (Sansevieria species) takes any exposure. **6.** Cactus and succulent collection can take full sun from the south and west in summer. **7.** Bulbs in containers moved indoors need no special light while in bloom. **8.** Budding azaleas brought indoors will bloom in direct eastern sun. **9.** Cut flowers from garden add color. **10.** Thanksgiving cactus (Zygocactus truncatus) needs filtered or diffused sunlight. **11.** Container chrysanthemums in bloom, brought indoors, need no special light. **12.** Poinsettias forced into bloom at Christmas. **13.** Jerusalem Cherry (Solanum pseudocapsicum) may be brought indoors while blooming at Christmas. **14.** Mini vegetable/herb garden in pots, trays, and hanging baskets, thrives in direct winter sunlight.

Starting your outdoor garden indoors

Flowers	Sowing Depth	Soil Temperature For Germination	Starting Time* (weeks before last killing frost)	Growing Tips
African daisy (Dimorphotheca)	1/8"	60-65 degrees	8-10	Important to germinate at cool soil temperature (below 70 degrees).
Ageratum	On surface	70-75 degrees	12-16	Fluorescent light will aid germination.
Alyssum	On surface	65-70 degrees	8-10	Water only with warm water.
Amaranthus	Cover lightly**	75-80 degrees	4-6	Keep soil temperature above 70 degrees for best germination.
Aster	1/8"-1/4"	65-70 degrees	6-8	Grow wilt resistant varieties.
Balsam	On surface	70-75 degrees	6-8	Best germination at a constant soil temperature of 70 degrees.
Begonia, fibrous	On surface	70-75 degrees	16-20	Use dilute liquid feed when seedlings visible.
Browallia	On surface	70 degrees	12-14	Light will aid germination.
Calendula (Pot marigold)	1/4"-1/2"	70 degrees	4-6	Keep seed covered with planting medium to exclude light.
Celosia (Cockscomb)	Cover lightly**	70 degrees	6-8	Seed is sensitive to drying out.
Coleus	On surface	70-75 degrees	10-12	Light needed for germination.
Dahlia	1/8"	65-70 degrees	8-12	Best germination with a constant 70 degree soil temperature.
Dianthus (annual)	Cover lightly**	70 degrees	6-12 (depending on variety)	Germinate perennial varieties at 75-80 degrees.
Gazania	1/8"-1/4"	60 degrees	10-12	Exclude light for best germination.
Geranium (Hybrid)	1/8"	70-75 degrees	10-12	Germination temperature is critical.
Heliotrope	1/8"	70-75 degrees	10-12	Uniform soil temperature best for germination.
Impatiens	On surface	70-75 degrees	12-16	Fluorescent light aids germination. Do not expose to direct sunlight.
Lobelia	On surface	70-75 degrees	12-14	Water with warm water. Variety 'Heavenly' requires 50 degrees for germination.
Marigold	1/4"	70-75 degrees	6-8	Dwarf marigolds are slower growing than taller varieties.
Nicotiana (Flowering tobacco)	On surface	70-75 degrees	6-8	Limit fertilizer to avoid leafy growth.
Petunia	On surface	70-80 degrees	10-12	Sow double varieties 2-4 weeks earlier.
Rudbeckia (perennial, Gloriosa daisy)	Cover lightly**	70-75 degrees	8-10	Will flower the first summer.
Salvia	On surface	70-75 degrees	8-10	Light aids germination.
Snapdragon	On surface	65-70 degrees	10-12	Pre-chill seed in refrigerator for a few days.
Verbena	1/4"	65 degrees	8-10	Germinate in dark. Water the sowing medium one day before sowing.
Vinca (Periwinkle)	1/8"	70-75 degrees	12-14	Germinate in the dark. Sensitive to over watering.
Zinnias	1/8"-1/4"	70-75 degrees	4-6	Sow as late as possible to avoid straggly growth.

Vegetables	Sowing Depth	Soil Temperature For Germination	Starting Time* (weeks before last killing frost)	Growing Tips
Broccoli	1/2"	65-70 degrees	4-6	Grow cool 55-60 degrees. Frost tolerant.
Brussels sprouts	1/2"	65-70 degrees	4-6	Grow cool 55-60 degrees. Frost tolerant.
Cabbage	1/2"	65-70 degrees	4-5	Grow cool 55-60 degrees. Frost tolerant.
Cauliflower	1/2"	65-70 degrees	5-6	Grow cool 55-60 degrees. Frost tolerant.
Cucumber	1"	70-75 degrees	3-5	Seed in a transplantable*** container. Frost sensitive.
Eggplant	1/4"-1/2"	70-75 degrees	6-8	Frost sensitive.
Lettuce	1/4"-1/2"	55-60 degrees	4-5	Grow cool 55-60 degrees.
Muskmelon	1"	75-80 degrees	3-5	Seed in a transplantable*** container. Frost sensitive.
Onion	1/2"	65-70 degrees	6-8	Very frost tolerant.
Pepper	1/4"-1/2"	75-80 degrees	6-8	Frost sensitive.
Tomato	1/2"	75-80 degrees	6-8	Frost sensitive.
Watermelon				Frost sensitive. Seed in a transplantable*** container.
Regular	1"	75-80 degrees	4-5	
Seedless	1"	75-80 degrees	5-6	

*Check with the local weather bureau or your County Agricultural Agent for the last spring frost date in your area.
**Lightly press the seeds into the medium surface.
***Such as peat pots.

NOTE: Thoroughly mist sown containers or moisten by placing them in water. Cover, after draining, with a polyethylene bag, plastic wrap, waxed paper, or with clear glass. Remove covering as soon as seeds have sprouted.

The man-made sun

Turn dark into light for healthier plants. Switch on to totally artificial light gardening or supplement natural sources.

◊

Various fluorescent lamps emit different "colors" as they add or subtract light rays. Top to bottom: Daylight, Deluxe warm white, warm white, cool white, Gro-Lux, Agro-Lite, and Vita-Lite.

So, you've made every effort to increase natural light. You've:

✔painted your rooms matte (flat) white;

✔invested in curtains and shades that open all the way;

✔trimmed or removed exterior foliage that was blocking light;

✔repositioned your plants so that they receive the maximum amount of light;

And still, that schefflera next to the sofa isn't growing and appears weak and spindly. You've fed and watered it properly, but it just isn't doing well. Not enough light. *It's too dark in the room.*

Or, you are a gesneriad fancier. Your friends have been growing prize African violets for years, while yours have produced few blooms. But the adjacent building blocks out most of the sun's radiant energy. *Not enough light* getting into the room.

Or, you want to get a jump on spring with garden seedlings started indoors. But because you live in Minnesota you are not as fortunate as Texans who enjoy longer and sunnier winter days and can make good use of a kitchen window garden. *Your* kitchen just *doesn't get enough light.*

Or, you've always enjoyed that spectacular view from your apartment—it's the toast of your friends. But having a northern exposure has limited the plants you may grow. There just isn't enough light to bring your cattleya orchid into flower, or to support a healthy weeping fig. *It's too dark.*

Turn dark into light

Artificial lighting can solve the above problems and many others. Plants can grow and bloom in a situation where artificial light is their sole energy source. Or, more naturally, artificial light can be used to supplement sunlight. When the day shortens in autumn, most plants will have a slower growth rate. However, artificial light can be employed to extend the growing season.

Any type of artificial light will help a plant, even if it is your floor lamp turned on for only a few hours each evening. Bright overhead lights will go a long way in making up light deficiency for tall plants. But in a room poorly lighted by the sun, ordinary house lighting will not be enough. Specialized lighting will have to be installed. It can make a garden of the dimmest room where no foliage type plants would ordinarily survive and will allow flowering plants to blossom in places that would normally support only foliage plants.

Some commercial flower growers have used artificial lighting for years to supplement production of florists' delights in quality and quantities that they would be unable to achieve in any other fashion. The bedding plant industry utilizes artificial lighting to start seedlings. And since the 1960's African violet enthusiasts and other light gardeners interested in the culture of specific plants have established often-elaborate setups in basements, attics, or in living room bookshelves.

Artificial light does not exactly duplicate sunlight. Its colors are present in different proportions. But certain types of artificial light can faithfully induce the natural responses in plants. Not all plants can be made to thrive under artificial light since this light is not nearly as intense as natural sunlight. However, a vast number of plants require only a moderate amount of sunlight in nature and can be made to perform quite well indoors under lights. As long as the required light waves are present in the proper proportions the source of the light is unimportant.

Since fluorescent lights are rich in foliage-producing blue rays, plants grown primarily for their foliage effect can subsist happily on fluorescents alone. Plants that flower, however, require red and far-red energy as well—sunlight used as a supplement, incandescent light, or a full spectrum fluorescent—if they are to perform well.

Artificial light is produced by a broad array of lamps. We are concerned here only with those fixtures and lamps that are adaptable to indoor gardening use.

Incandescent light

The most common incandescent bulbs are the bulbs used everyday in the home. These bulbs consist of a tungsten filament wire which has high resistance to electricity. The resistance caused by the filament results in the emission of visible light.

Incandescents are rich in red and far-red light which are indispensable for flowering and other plant processes. In fact, incandescent light possesses the same proportion of red and far-red rays as sunlight, although vastly less intense. However, the total energy output is insufficient among the blue and violet rays of the spectrum and therefore the incandescent light as a sole energy source is not suitable for complete plant growth.

It can be used to supplement daylight when the light deficiency is minimal. Boston fern or spathyphyllum on a side table can get needed supplementary light from an ordinary 75-watt lamp burning three or four hours a night.

Incandescents also give off a considerable amount of heat. This heat can be damaging to plants if the light source is placed too close. A rule of thumb is: *If your hand feels warm when held at the foliage closest to the light source, the plant is too close. Generally speaking, it is best to keep the tops of plants at least a foot away from incandescent sources.* Naturally the lamp should not be placed too far away from the plant either.

The light reaching the plant will decrease with the square of the distance that it is removed. That is, a plant two feet away from a light source will receive only one-fourth as much light as it would if it were one foot away.

An easy way to reduce heat is to utilize several smaller bulbs instead of one large one. This distributes the heat over a large area, allows placing the lamps closer to the plants and provides more even distribution of light. If this is impractical a shield of glass or transparent plastic will absorb or reflect a large amount of heat while allowing nearly all of the light to pass through to the plant. The shield should be placed several inches away from the lamp.

Another method of avoiding heat problems is to use reflectorized incandescent lamps such as Cool Beam or Cool Lux. These bulbs contain a silverized reflecting surface which reflects light downward but conducts heat upward. The heat projected downward by this method may be reduced 50% or more. A word of warning: These bulbs should always be used in a ceramic socket, as a regular socket will burn out.

Incandescent "plant growth" lamp provides main light source for a corner bedroom garden.

Fluorescent light

Fluorescent lamps are familiar to all of us as the ones commonly used in offices, factories and public places. Introduced in 1938, these lamps have enjoyed success because of their low cost, long life and evenness of light distribution. They offer higher light efficiency than incandescent lamps.

A fluorescent lamp of the same wattage as an incandescent lamp emits 2½ to 3 times as much light. Fluorescents generate only a very small amount of heat. The ballast, which is a part of the lamp fixtures, emits heat, but in most fixtures is far enough removed from the bulb that plants can actually touch a fluorescent bulb with no adverse effects. The lifetime of a fluorescent tube is about 15-20 times that of an incandescent bulb. Thus, in the long run, it is much more economical.

The long tubular glass bulb of fluorescent lamps is coated on the inside with a phosphor. It is the type of phosphor which determines the 'color' of the light given off. It is the mixture of phosphorescent chemicals that determine the 'mix' of the various color wavelengths. *The visible color, however, is not indicative of the proportion of blue and red waves given off.* The bulb contains a blend of inert gasses such as argon, neon, or krypton, and a minute quantity of mercury vapor, sealed in a low pressure. There are two electrodes, one at each end of the tube. Electricity induces a current to flow between the electrodes in the form of an electrical arc which stimulates the phosphor and emits energy in the form of light. *This emission is stronger in the center of the tube than at the ends.*

Electricity is regulated by ballasts, which are small transformers that reduce the current in the tube to the operating voltage required by that particular lamp. The ballast is usually contained in the fixture and emits most of the small amount of heat given off by a fluorescent lamp. In the most common home-lighting circumstances the heat given off by ballasts is not a problem unless a great number of large fixtures are used in one small area where the overall room temperature would be raised. If an individual fixture seems to be radiating too much heat downward, the ballast can be removed from the metal tray that houses the ballast and a piece of asbestos placed between the ballast and tray.

However, some set-ups require the installation of a large number of lamps. In these cases, the ballasts can be installed away from the lamps themselves. Normally this should be done by an experienced electrician.

Ballasts use about ten watts of electricity for every 40 watts used by the lamp itself. This is important to figure in when large installations are being

This begonia collection thrives and blooms with fluorescents as the only light source.

Fluorescent tubes hidden underneath the window valance are used to supplement sunlight in Clyde Childress' hanging tomato garden.

contemplated so that circuits are not overloaded. Ballasts normally last ten to twelve years. When they burn out they sometimes smoke and give off a noxious chemical odor. They can be easily replaced without professional help by the average homeowner; check with your local electrical or hardware dealer.

Fluorescent tubes are held in the fixtures by pins at both ends of the tube. There are four types of pins:

✓2-pin (Medium Bi-pin) is the most common, usually found on bulbs with lower wattages, up to 90 w.

✓1-pin on narrower diameter bulbs and on the intermediate wattages.

✓Recessed Double Contact on higher wattage bulbs.

✓4-pins on circular bulbs of various wattages.

Fluorescent lamps don't 'turn on' the way incandescent lamps do. The bulb's cathodes must be warm before an arc can be struck through the lamp. The starting process is a major factor in bulb wear. A pair of 40 watt Cool White fluorescent bulbs will last about twice as long if burned continuously than if burned three hours a day, because of the energy and wear used in the starting process.

Fluorescent lamps are available in various colors:

White	*Soft White*
Cool White	*Deluxe Warm White*
Warm White	*Deluxe Cool White*
Daylight	*Supermarket White*
Sign White	*Merchandising White*
Living White	

'Color' refers to the quality of the light given off, not the bulb's temperature. The Cool White tube is the traditional lamp of the light gardener. Because most fluorescent light is not as attractive to the eye as incandescent light, many people leave them on during the daytime when natural daylight can help offset the harshness of the light or when they are away at work. This adds only a few cents per day to the electrical bill. The Deluxe Warm White is the most flattering fluorescent to complexions and home furnishings.

There is a bewildering array of lamps available on the market. Sylvania, General Electric and Westinghouse are the largest manufacturers of fluorescent bulbs. They list a full range of sizes, shapes, colors and wattages in their catalogs. The chart on page 87 goes into some detail with representative bulbs from major manufacturers for the benefit of the interior gardener.

Window shades can be lowered to serve as a reflector for light from "plant-growth lamp". This is useful when plants need a period of increased light intensity. For additional information and photos on plant cart, see page 16.

Although initially more expensive than incandescent bulbs, fluorescent lamps form the backbone of light gardening. As we have seen, they throw off so little heat they can be placed very close to the plants. They can be positioned as close as one inch to many blooming plants, although six to nine inches is more common. Nevertheless, they should not be placed more than 18 inches away from the top of a plant with any expectation of flowering, unless the lamps are higher output. For best flowering they should operate 10 to 18 hours a day if they are the sole light source. Timers are a great help in keeping the light schedule, particularly in allowing regularity when one is away from home. Timers should be fully adjustable to any period over a full 24 hours and should be of adequate capacity to carry the wattage demanded. Low cost timers are available in most hardware stores. Many people burn their lights regularly on timers when they're away to discourage burglars, so the light may as well be benefiting the plants at the same time.

Because fluorescents, like incandescents, 'blacken' with age and lose light efficiency, it is recommended by many that bulbs be replaced when they reach 70% of their stated service life. (Usually listed on the bulb package or label.) At this point they provide about 15% less light than when new. A grease pencil entry on the end of the tube marking the installation date will assist in timely replacement. A 'flickering' lamp is about to burn out and should be replaced. Light gardeners with banks of lamps often stagger the lamp replacement to prevent 'light shock' to plants accustomed to dimming light.

There are only three methods of increasing fluorescent light: Add more fixtures, reflect more light on the plants, or move the plants closer to the lamps. Using longer tubes will increase light *because there is considerable light loss at the ends of tubes.* There is twice as much light one foot away from a fluorescent light source than there is two feet away. Fluorescent tubes can be dimmed by rheostats only if special dimming ballasts are installed. These are available for use with the 30 and 40 watt tubes.

Ordinary fluorescent light in the right intensities can promote lush foliage growth, even if there is not a drop of sunlight available. But, like incandescent light, it's short on certain parts of the spectrum needed for entirely-balanced plant life. All common fluorescent output is very high in blue light, which promotes foliage growth and is extremely low in red and far-red light, which promotes flowering.

Two ways for proper light proportions

There are two ways in which the above dilemma may be resolved while still by-passing the sun and providing light in the proper proportions to support all plant processes:

1. Combine fluorescent and incandescent light.
2. Use wide spectrum 'plant growth' lamps.

THE FIRST METHOD has been, until recently, the only alternative. Light gardeners have used it for some time, especially for flowering and fruiting, and fixtures are available that accommodate both types of bulbs in the same unit. They are somewhat costly and use a greater amount of electrical energy than the common fixture.

We found so many varying ratios for the fluorescent/incandescent mix that it is difficult to do more than report some of them. One incandescent watt to each five fluorescent watts has been suggested, as has 1:4, 1:3 and 1:2. One light gardener suggests a 1:3 ratio for combined wattages below 1,000 f.c. and increasing it to 1:2

Mark the end of fluorescent tubes with a grease pencil to indicate installation date. This helps you to know when replacement time is approaching.

above 2,000 f.c. There are also varying formulas for combining particular lamps for best effect. One Warm White type might be combined with one Cool White type, or two Cool Whites to one Deluxe Warm White. Or, reflectorized incandescents might provide the red, while Cool White comes up with the blue!

THE SECOND AND EASIEST METHOD is to use wide spectrum 'plant growth' lamps. These bulbs come close to providing all of the wavelengths in natural daylight needed for plant health and providing them in the proper proportions. Here, then, is sunlight in a tube!

"Plant-growth" and "wide-spectrum" lamps

These are modified fluorescent lamps and are sold under the trade-mark names of Gro-Lux, Plant-Gro and Plant-Light, among others. Since the green or yellow parts of the spectrum have no known influence on a plant's biological functions, special fluorescent phosphor lamps were developed which minimize these rays and concentrate instead on the emission of blue and red rays needed for healthy plant growth. Because of the missing values the light given out tends to be purplish or pinkish. This light enhances flower color—reds scream red, pinks become phosphorescent and yellows glow. Foliage appears lusher under "grow lights"—a richer dark green than they do under 'white' light. This special effect can be exploited in interior decorating, although the intense purplish effect of some lamps may be difficult to deal with in some situations. If you find the color offensive, use the special lamps only when the room is not in use and provide alternate lighting when you're at home.

The newer 'wide spectrum' growth tubes include the visible blue and far-red rays, and sometimes ultra-violet rays as well. They not only aid in plant growth, but provide a more realistic visual rendition of plant color than the earlier plant lights. Gro-Lux Wide Spectrum has a pink colored light which is easier to accommodate to the home environment than some of the darker hues. Due to the added 'far-red' in wide spectrum bulbs, which is not visible, the light given off tends to appear about 78% as strong as that emitted by cool white tubes. Trade names for the wide spectrum bulbs are called Gro-Lux Wide Spectrum, Vita-Lite, and Naturescent/Optima, among others.

Bulbs for light gardening

See diagrams on next two pages.

Type	Wattages	Hours	Length	Diameter	Pins
Incandescent					
General Household Lamps	15, 25, 40, 50, 60, 75, 100, 150, 200, 500	750-1,000			
Reflector Floodlights and Spotlights	30, 50, 75, 150	1,000			
Parabolic Reflectorized (PAR)	75, 150	1,000			
Cool Beam Flood Lamps	75, 150, 300	1,000			
PS-30 Reflectorized Bulbs (PS)	150, 250	1,000			
Lumiline Tubes	30, 40, 60	1,000			
Showcase Bulbs	25-75	1,000			
Plant Life	75, 150	1,000			
Fluorescent					
Standard Preheat	4-90	6,000-22,500	6″-60″	⅝″-2⅛″	bi-pin
Instant-Start	40	7,500	48″	1½″, 2⅛″	bi-pin
Rapid-Start	30, 34, 40	18,000-34,000	36″, 48″	1½″	bi-pin
Instant-Start Slim Line	20-75	7,500-22,500	42″-96″	¾″-1½″	single pin
Rapid-Start High Output (HO)	25-110	9,000-22,500	18″-96″	1½″	recessed double contact
Rapid-Start Very High Output (VHO) and Rapid Start Super High Output (SHO)	110-219	9,000-16,000	48″, 72″ 96″	1½″	recessed double contact
Reflectorized T-12	110, 135, 165, 215	12,000	48″, 96″	1½″	recessed double contact
Power Groove and Power Twist	110, 160, 215	9,000	48″, 60″ 72″, 96″	2⅛″	recessed double contact
Mod-U-Line, U-Bent, Curvalume	40	12,000	24″ x 6″ 24″ x 3⅝″	1½″	bi-pin
Circline	22, 32, 40	9,000	8¼″, 12″, 16″	1⅛″	4-pin
Plant-growth					
Gro-Lux	8, 14, 15, 20, 30, 40	6,000-20,000	15″, 18″, 24″, 36″, 48″	1″, 1½″	bi-pin
	73	12,000	96″	1½″	single pin
	105, 110, 160, 215	10,000-12,000	48″, 72″, 96″	1½″	recessed double contact
Plant-Gro					
Plant-Light	8, 14, 15, 20, 30, 40	6,000-12,000	12″-48″	⅝″, 1″, 1½″	bi-pin
Agro Lite	15, 20, 40	7,500-20,000	18″, 24″, 48″	1″, 1½″	bi-pin
Gro-Lux Wide Spectrum	40	7,500	48″	1½″	bi-pin
	75	7,500	96″	1½″	single pin
	105, 110, 160, 215	7,500	48″, 72″, 96″	1½″	recessed double contact
Vita Lite	14, 15, 20, 30, 40	13,000-33,000	12″-48″	⅝″, 1″, 1½″	bi-pin
	38, 51, 55, 75	13,000-22,000	36″, 48″, 72″	1½″	single pin
	60, 85, 110, 160, 215	10,000-14,000	72″, 96″	1½″	recessed double contact

Incandescent lamps

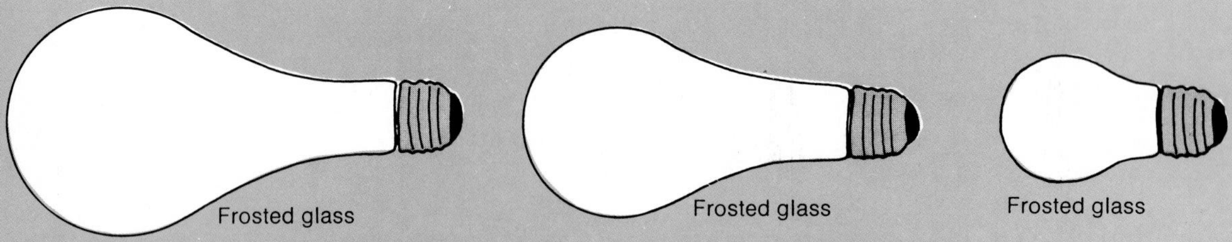

Frosted glass
Frosted glass
Frosted glass

General household lamps—150, 100, and 40 watts.

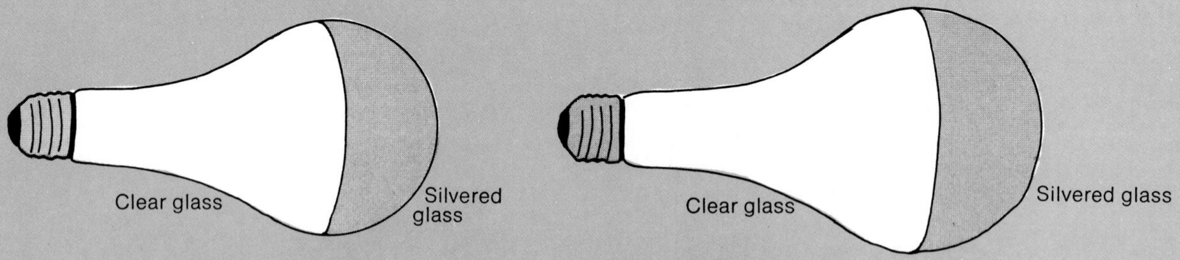

Clear glass
Silvered glass
Clear glass
Silvered glass

Silvered bowl lamps—100 and 150 watt

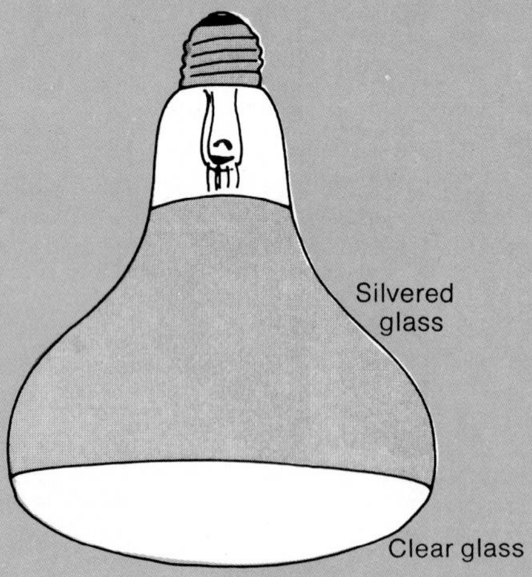

Silvered glass
Clear glass

Reflector flood lamp

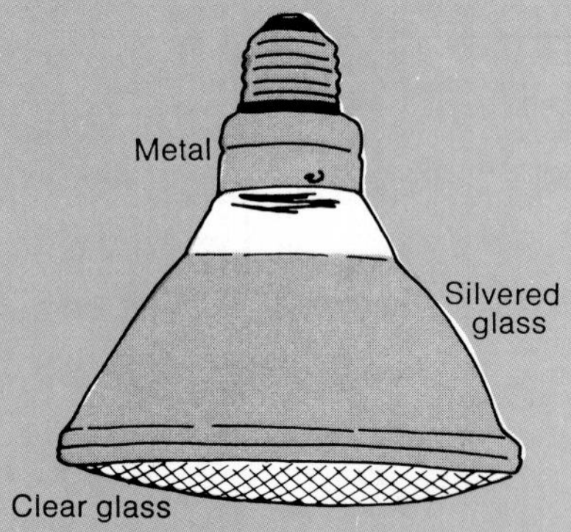

Metal
Silvered glass
Clear glass

Parabolic Reflectorized

Lumiline lamp

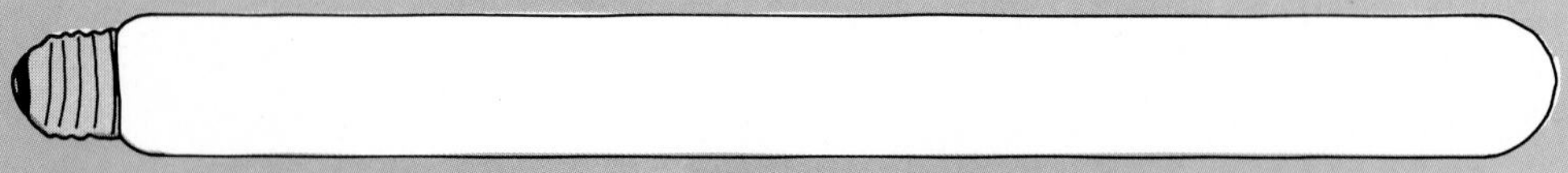

Showcase lamp

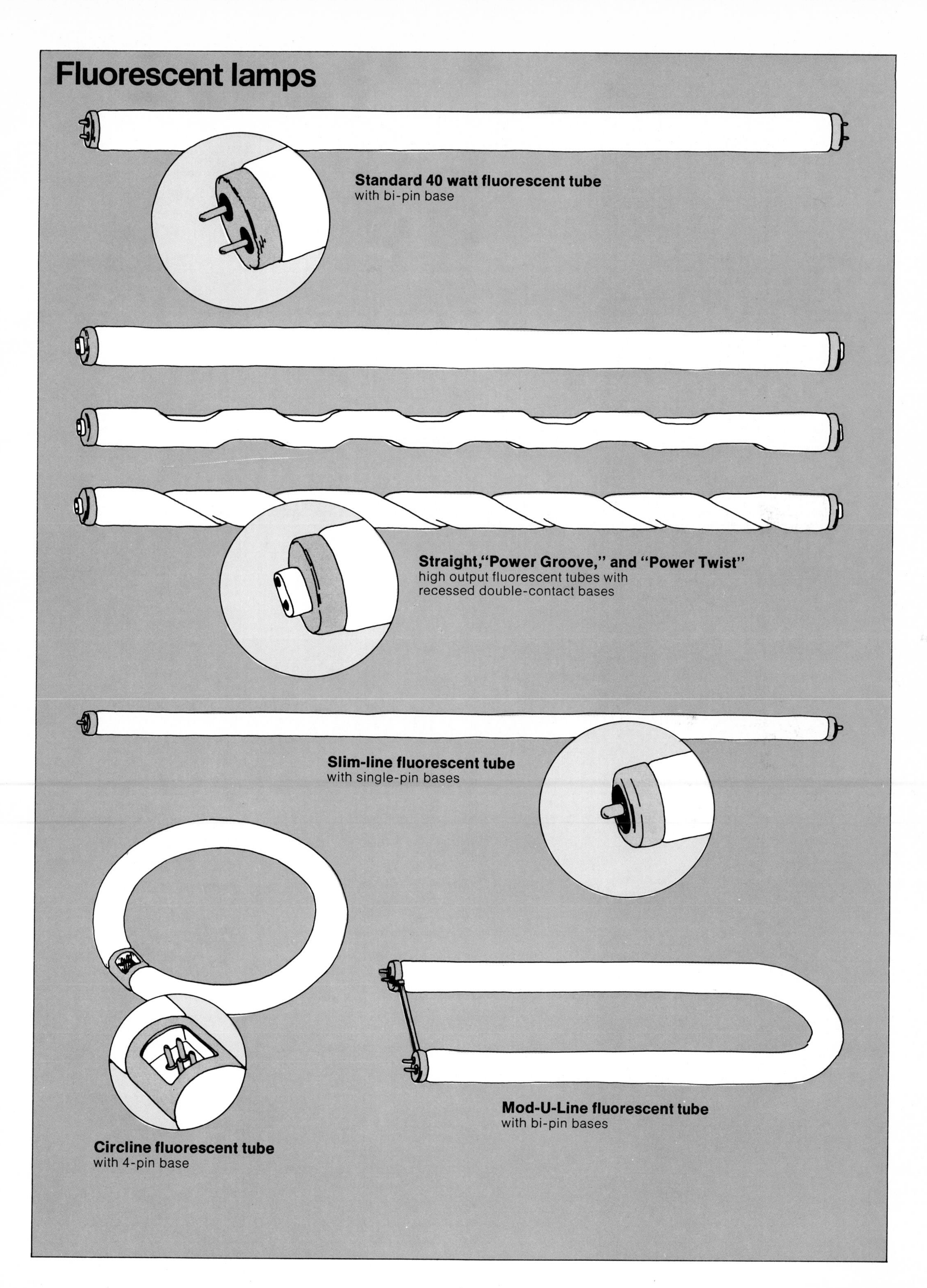
Fluorescent lamps
Standard 40 watt fluorescent tube
with bi-pin base
Straight,"Power Groove," and "Power Twist"
high output fluorescent tubes with
recessed double-contact bases
Slim-line fluorescent tube
with single-pin bases
Circline fluorescent tube
with 4-pin base
Mod-U-Line fluorescent tube
with bi-pin bases

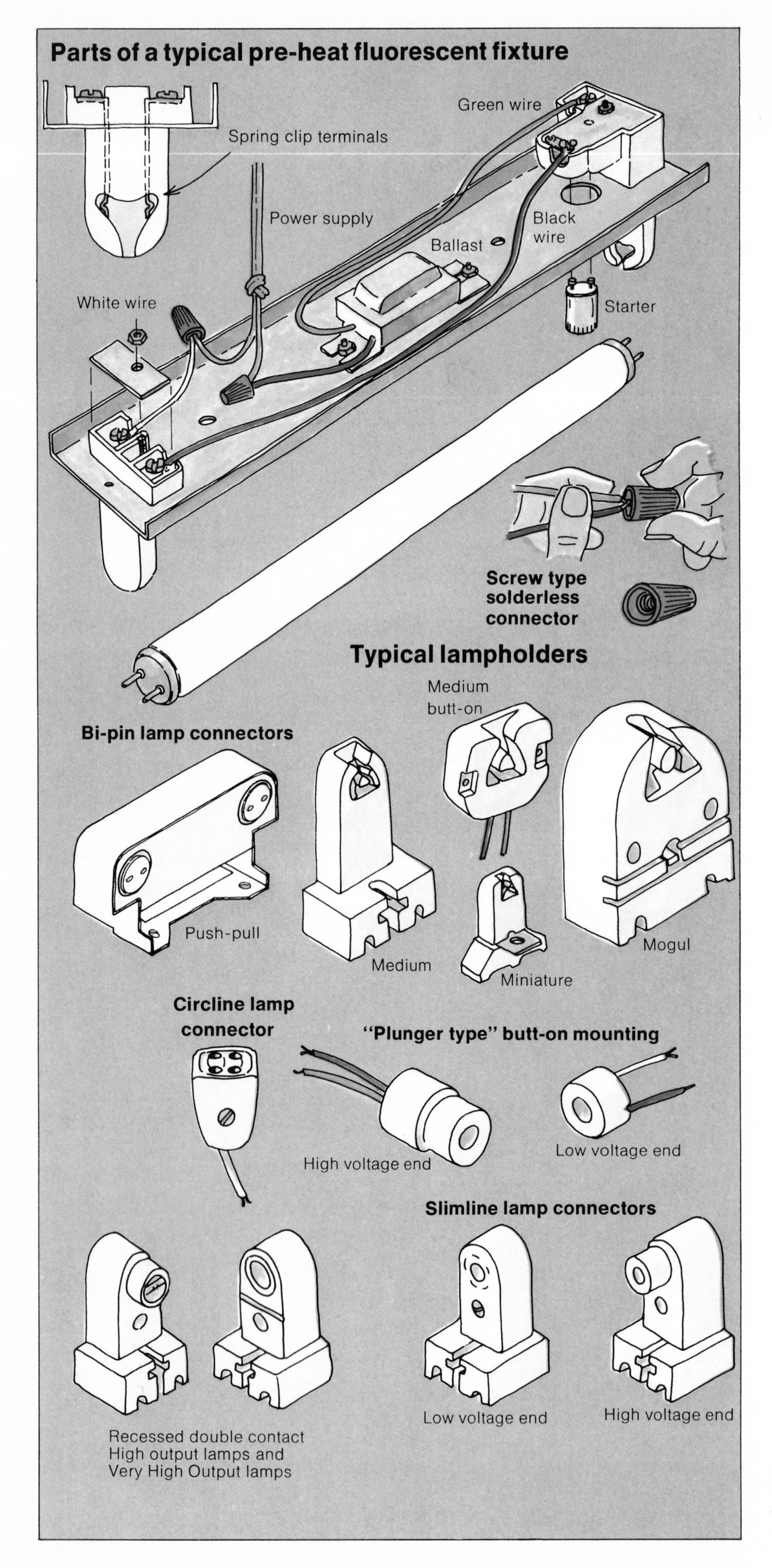

Light gardening fixtures and equipment

Incandescent sockets are the familiar screw-in type. They should be of the porcelain type when lamps larger than 75 watts are used to insure good contact and prevent shorting out.

Fluorescent tubes require their own special fixture. The two most common types of fixtures are 'industrial,' which have a built-in reflector and 'strip' or 'channel' which do not. The 'industrial' type, with reflector, is hung in the open where there would be no other means of reflecting light. Light gardeners with large setups most commonly use banks of the industrial type fixtures with reflectors.

The channel type is used in constricted places, like bookcases or cabinets, where the supporting background is made reflective by painting it white. (Flat, or matte white with a textured surface reflects more light than glossy paint with a smooth surface.)

Both fixture types are available in units that can accommodate from one to four tubes.

Lighting fixtures may be installed so that the light source is concealed or obscured. This may be particularly appropriate in the living or dining room where an exposed fixture would have a negative impact on the decor. The obvious disadvantage of this method is that the distance from lamp to plant is fixed and flexibility must then be achieved only through the use of bulbs of higher or lower wattages.

Temporary clamp-on incandescent sockets with reflectors or a photographer's lamp on a tripod allow for removal from the living room 'when company comes,' or for semi-permanent supplementary illumination during the winter months when the sun is south of the border.

Fluorescent fixtures may be suspended from chains in the basement installation to allow for adjustment necessitated by plant growth.

For the more casual light gardener, not interested in setting up elaborate basement installations, there are many plant stands, 'gro-light' furniture pieces and carts on the market that have two or three trays with fixtures, reflectors and automatic timers built in. Such items are avaliable from the accompanying list of plant lighting manufacturers.

Light gardening suppliers

✔ALADDIN INDUSTRIES, INC.
Horticultural Products
P.O. Box 10666
Nashville, TN 37210
Environmental cases for light gardens.

✔W. ATLEE BURPEE CO.
Warminster, PA 18974
Bulbs and equipment for light gardens.

✔CRAFT-HOUSE MANUFACTURING COMPANY
Wilson, NY 10706
Lighted plant stands.

✔DURO-LITE LAMPS, INC.
Duro-Lite Dept. PA-12
Fair Lawn, NJ 07410
Plant growth lamps.

✔DURO-TEST CORPORATION
North Bergen, NJ 07047
Fluorescent and incandescent bulbs.

✔ENVIRONMENT ONE
2773 Balltown Road
Schenectady, NY 12309
Fluorescent lighted growth chambers.

✔FLECO INDUSTRIES
3347 Halifax Street
Dallas, TX 75247
Fluorescent "Plant lighters" and incandescent "Grolighters"; and portable stands.

✔FLORALITE CO.
4124 E. Oakwood Road
Oak Creek, WI 53154
Fluorescent plant stands, fixtures and "plant growth" bulbs.

✔FLUORESCENT TUBE SERVICE
13107 South Broadway
Los Angeles, CA 90061
Bulbs, lamps and fixtures

✔GARCY CORPORATION
Spacemaster Home Products Division
2501 N. Elston Avenue
Chicago, IL 60647
Adjustable plant display trays and fluorescent fixtures.

✔GENERAL ELECTRIC COMPANY
Lamp Division
Nela Park
Cleveland, OH 44112
Incandescent and fluorescent lamps.

✔THE GREEN HOUSE
9515 Flower Street
Bellflower, CA 90706
Bulbs, stands and fixtures.

✔GROWER'S SUPPLY CO.
Box 1132
Ann Arbor, MI 48106
Lighting stands, trays, timers and tubes.

✔HALL INDUSTRIES
2323 Commonwealth Avenue
North Chicago, IL 60064
Circular fluorescent tubes in decorative growth lamps.

✔HOUSE PLANT CORNER
P.O. Box 810
Oxford, MD 21654
Lighted plant stands.

✔H. L. HUBBELL INC.
Zeeland, MI 49464
"Furniture that grows" in traditional and contemporary styles.

✔LIFELITE
1036 Ashby Avenue
Berkeley, CA 94710
Decorative fluorescent units and light garden supplies.

✔LORD & BURNHAM
Irvington, NY 10533
Enclosed fluorescent light garden units.

✔MARKO
94 Porete Avenue
N. Arlington, NJ 07032
Units and other equipment.

✔NEAS GROWERS SUPPLY
P.O. Box 8773
Greenville, SC 29604
Light garden table top units.

✔GEORGE W. PARK SEED CO. INC.
Greenwood, SC 29646
Lamps and other equipment.

✔SERVICE LIGHTING PRODUCTS
Capitol Street
Saddle Brook, NJ 07662
Plant stands and carts with Gro-Lites.

✔SHOPLITE CO.
566 Franklin Avenue
Nutley, NJ 07110
Specialists in hard-to-find items.

✔H. P. SUPPLIES
16337 Wayne Road
Livonia, MI 48154
Fluorescent fixtures, bulbs and tubes.

✔SYLVANIA ELECTRIC PRODUCTS, INC.
60 Boston Street
Salem, MA 01971
All types of fluorescent lamps.

✔TUBE CRAFT, INC.
1311 West 80th Street
Cleveland, OH 44102
Light units, trays and timers

✔WESTINGHOUSE ELECTRIC CORPORATION
Westinghouse Lamp Division
Bloomfield, NJ 07003
All types of fluorescent lamps.

✔VEND-A-RAY CORPORATION
615 Front Street
Toledo, OH 43605
"Plant lights" and fixtures.

Left: "Plant lighter" étagère from Fleco Industries comes with "plant-growth lamps" built into the shelves. Upper right: A "flora cart" is used to display plants growing in a dark basement. Lower right: Tabletop units shed light on a collection of small plants.

Vita-lite fluorescents on automatic timers provide the sole light source for modular garden compartments covering a wall of a New York City townhouse.

An inverted clay pot inside a macrame hanger is used as a lamp shade for a "plant-growth" incandescent to light a spider plant. Several lighted hanging plants may provide all the evening illumination needed in a room.

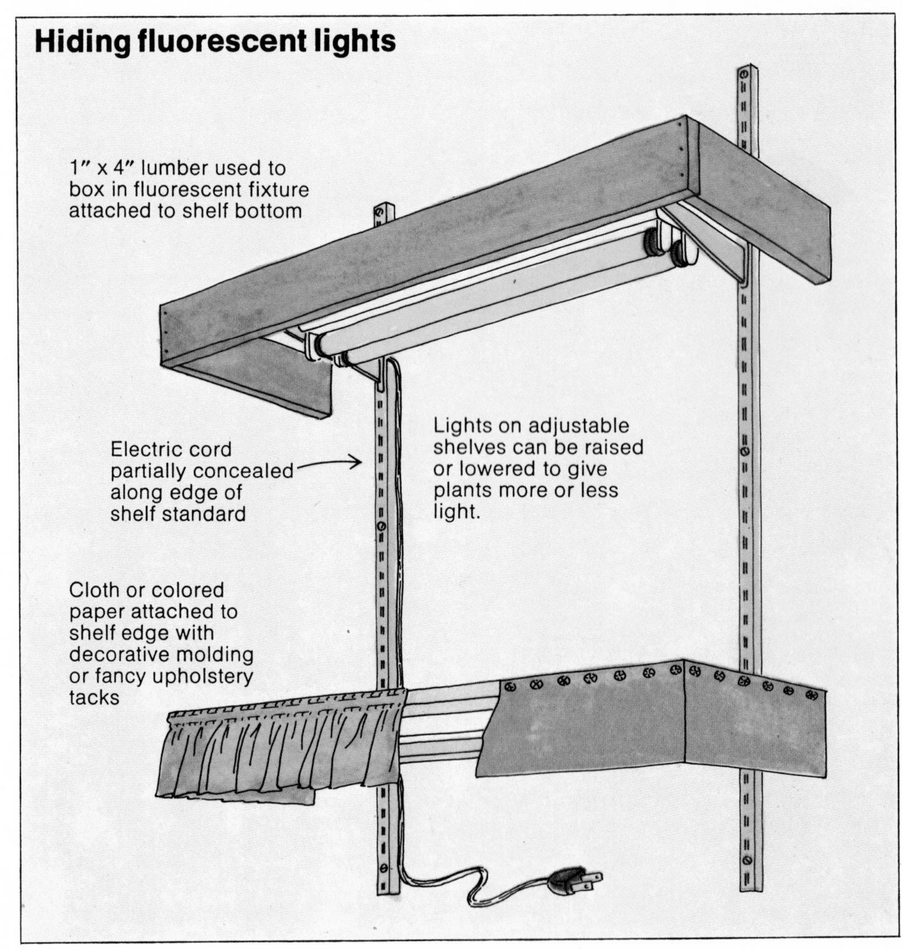

We built a mobile lighted case that can be used against the wall or floated in the room as a divider. See instructions for construction on right.

The fireplace can be a summer garden. Here "plant-growth" fluorescents are installed inside the chimney. Mylar-covered panels cover sides and back for extra reflectance. A glass or plastic screen can be added over the opening for a terrarium-like atmosphere.

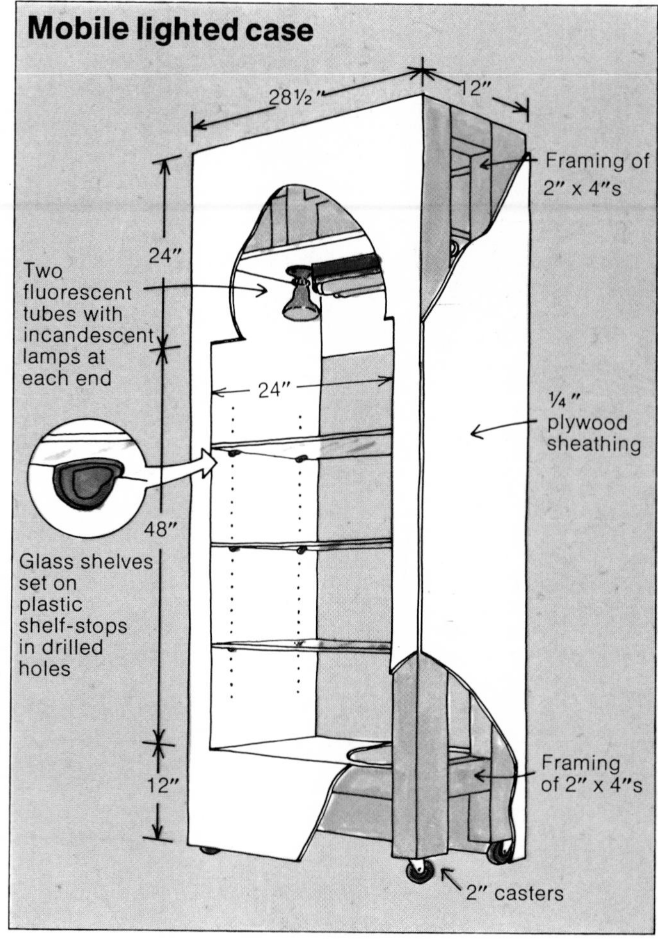

Plant display

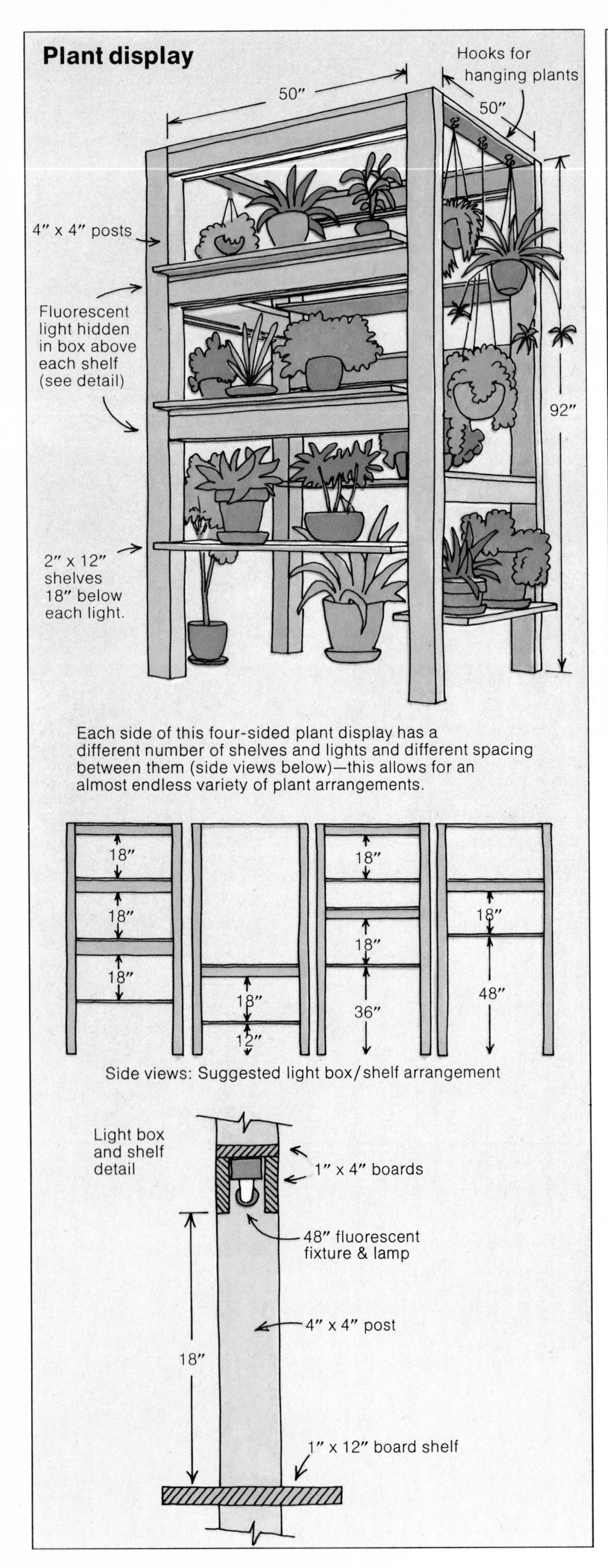

Light-box atrium

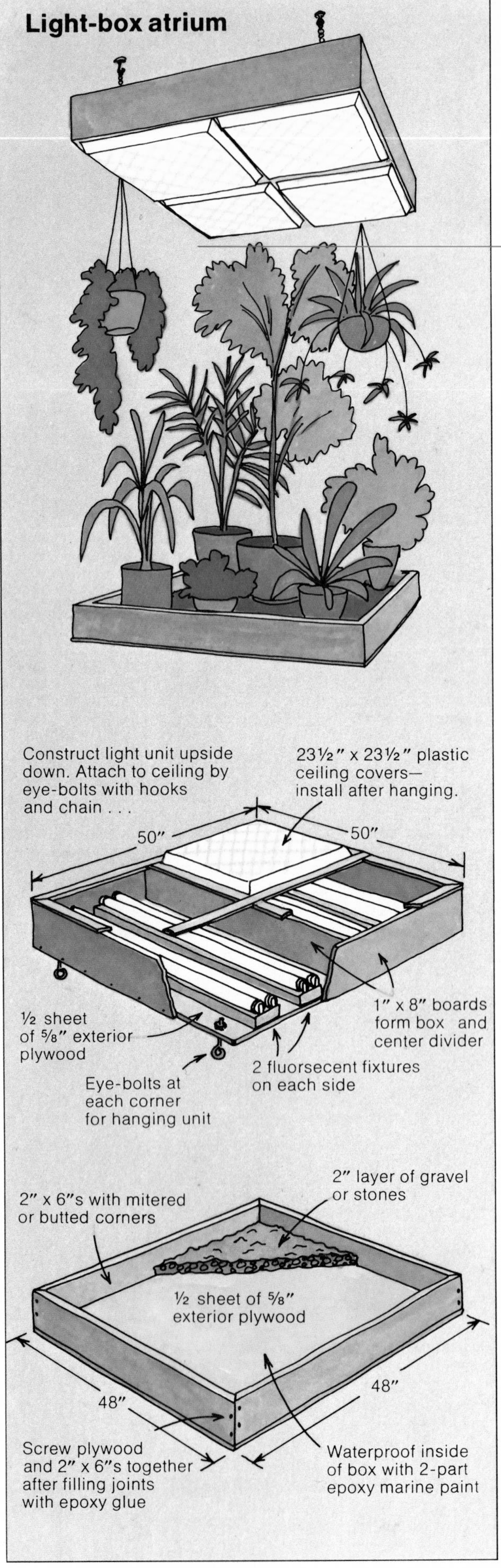

Clamp-on incandescent sockets with reflectors provide temporary "plant growth" light. Low light plants can thrive under such light.

A photographer's tripod sheds more light on plants when needed, then it can be folded up and stored away "when company comes."

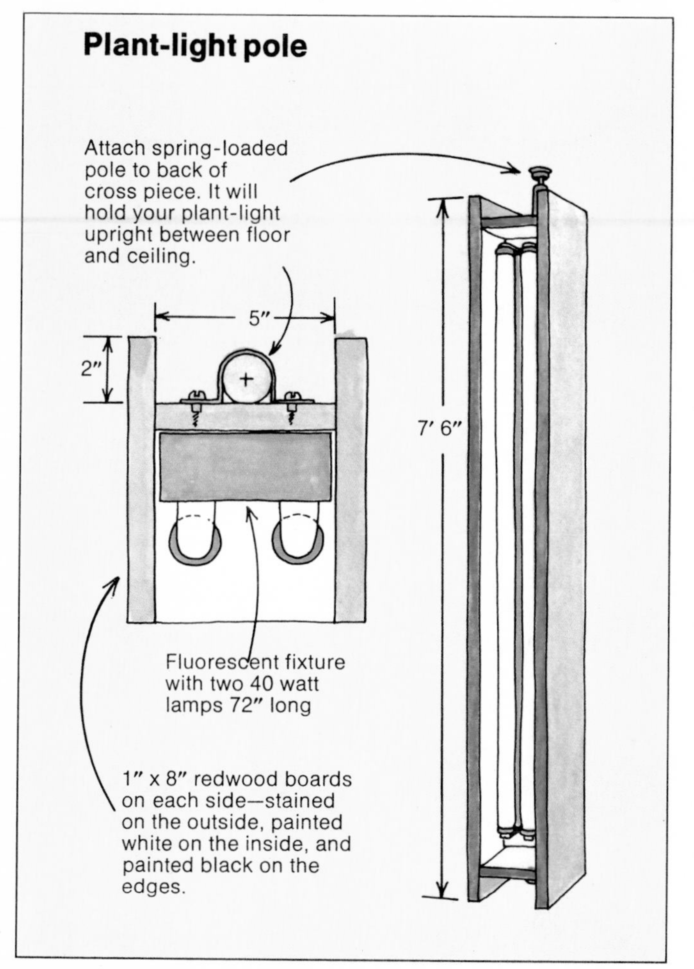

Vertical "plant-growth" fixture for tall plants is easily constructed. Finish sides to fit your interior design scheme.

Help for indoor gardeners

Books

BRINGING THE OUTDOORS IN
H. Peter Loewer
Walker and Company

EXOTICA
Alfred Byrd Graf
Roehrs Company, Inc.

FLOWERING HOUSE PLANTS AND FOLIAGE HOUSE PLANTS
James Underwood Crocker
Time-Life Encyclopedia of Gardening

GARDENING IN A BOWL
Elvin McDonald and James McNair
Doubleday

GARDENING INDOORS UNDER LIGHTS
Frederick H. & Jacqueline L. Kranz
Viking Press

HOME ORCHID GROWING
Rebecca Tyson Northern
Van Nostrand Reinhold Company

HOUSEPLANTS FOR CITY DWELLERS
Alys Sutcliffe
E. P. Dutton

HOUSE PLANTS FOR THE PURPLE THUMB
Maggie Baylis
101 Productions

HOUSEPLANTS INDOORS/OUTDOORS
Edited by Elvin McDonald & James McNair
Ortho Book Series

HOW TO GROW BEAUTIFUL HOUSE PLANTS
T. H. Everett
Fawcett

INDOOR PLANT SELECTION AND SURVIVAL GUIDE
Terrestris
Grosset & Dunlap

LIGHTING FOR PLANT GROWTH
Elwood D. Bickford & Stuart Dunn
The Kent State University Press

MAKING THINGS GROW
Thalassa Cruso
Knopf

THE AVANT GARDENER
Thomas and Betty Powell
Houghton Mifflin Company

THE COMPLETE BOOK OF GARDENING UNDER LIGHTS
Elvin McDonald
Doubleday

THE COMPLETE BOOK OF HOUSEPLANTS
Charles Marden Fitch
Hawthorn

THE COMPLETE BOOK OF HOUSEPLANTS UNDER LIGHTS
Charles Marden Fitch
Hawthorn

THE COMPLETE INDOOR GARDENER
Edited by Michael Wright
Random House

THE HOUSE PLANT ANSWER BOOK
Elvin McDonald
Popular Library

THE WORLD BOOK OF HOUSE PLANTS
Elvin McDonald
World Publishing and Popular Library

THE INDOOR LIGHT GARDENING BOOK
George A. Elbert
Crown Publishers

Magazines

FLOWER AND GARDEN
4251 Pennsylvania
Kansas City, MO 64111
$3.00 per year—monthly
Regionalized for Northern, Southern, and Western. All aspects of home gardening.

HORTICULTURE
125 Garden Street, Marion, OH 43302
$8.00 per year—monthly
Colorful—all phases of horticulture.

ORGANIC GARDENING & FARMING
Emmaus, PA 18099
$6.85 per year—monthly
All about organic gardening
Some help for indoor gardener.

PLANTS ALIVE
1255 Portland Place, Boulder, CO 80302
$9.00 annually—monthly except July and August
Good information for indoor and greenhouse gardeners.

THE AVANT GARDENER
P.O. Box 489
New York, N.Y. 10028
$10.00 per year—bi-weekly
Digest of latest in horticultural events, discoveries, products and news.

UNDER GLASS
Lord & Burnham, P.O. Box 114, Irvington, NY 10533
$2.50 per year—bi-monthly
Interesting for all indoor gardeners, especially for owners of greenhouses.

Plant societies and publications

These societies change officers regularly which often creates a change in mailing addresses. At the time of publication the following addresses were current.

AFRICAN VIOLET SOCIETY OF AMERICA, INC.
Box 1326, Knoxville, TN 37901
Membership $6.00 yearly includes African Violet Magazine *5 times per year.*

AMERICAN BEGONIA SOCIETY, INC.
1431 Coronado Terrace, Los Angeles, CA 90026
Membership $4.00 per year includes
The Begonia *monthly.*

THE AMERICAN BONSAI SOCIETY
229 North Shore Drive, Lake Waukomis, Parksville, MO 64151
Membership $10.00 per year includes
Bonsai *quarterly.*

AMERICAN FERN SOCIETY
Department of Botany, University of Tennessee
Knoxville, TN 37916
Membership $5.00 per year includes American Fern Journal *quarterly.*

THE AMERICAN FUCHSIA SOCIETY
738-22nd Avenue, San Francisco, CA 94121
Membership $3.00 per year includes American Fuchsia Society Bulletin *monthly.*

THE AMERICAN GLOXINIA AND GESNERIAD SOCIETY, INC.
P.O. Box 174, New Milford, CT 06776
Membership $5.00 per year includes
The Gloxinian *bi-monthly.*

AMERICAN HORTICULTURAL SOCIETY
Mount Vernon, VA 22121
Membership $15.00 per year includes newsletter and quarterly magazine.

THE AMERICAN PRIMROSE SOCIETY
14015 84th Avenue, NE Bothell, WA 98011
Membership of $5.00 per year includes Quarterly of the American Primrose Society.

CACTUS AND SUCCULENT SOCIETY OF AMERICA, INC.
Box 167, Reseda, CA 91335
Membership $10.00 per year includes Cactus and Succulent Journal *bi-monthly.*

EPIPHYLLUM SOCIETY OF AMERICA
218 E. Graystone Avenue, Monrovia, CA 91016
Membership $3.00 per year includes
Epiphyllum Bulletin *irregular.*

THE AMERICAN ORCHID SOCIETY
Botanical Museum of Harvard University, Cambridge, MA 02138
Membership $10.00 per year includes American Orchid Society Bulletin *monthly.*

THE AMERICAN PLANT LIFE SOCIETY AND THE AMERICAN AMARYLLIS SOCIETY
Box 150, LaJolla, CA 92037
Membership $5.00 per year includes Plant Life—Amaryllis Yearbook *bulletin.*

THE INDOOR LIGHT GARDENING SOCIETY OF AMERICA, INC.
128 W. 58th Street, New York NY 10019
Membership $5.00 per year includes
Light Garden *bi-monthly.*

THE PALM SOCIETY
7229 SW 54th Avenue, Miami, FL 33143
Membership $10.00 per year includes
Principes *quarterly.*

SAINTPAULIA INTERNATIONAL
Box 10604, Knoxville, TN 37914
Membership $4.00 per year includes
Gesneriad Saintpaulia New *bi-monthly.*

Plant sources

ABBEY GARDEN
170 Toro Canyon Road, Carpinteria, CA 93013
Cacti and other succulents. Catalog $1.00.

ALBERTS & MERKEL BROTHERS
P.O. Box 537, Boynton Beach, FL 33435
Orchids and other tropicals. Catalog $1.00.

ANTONELLI BROTHERS
2545 Capitola Road, Santa Cruz, CA 95010
Tuberous begonias, gloxinias, achimenes.

ASHCROFT ORCHIDS
19062 Ballinger Way N.E., Seattle, WA 98155
Botanical orchids.

ALBERT H. BUELL
Eastford, CT 06242
African violets, gloxinias, other gesneriads. Catalog $1.00.

BURGESS SEED AND PLANT COMPANY
Box 2000, Galesburg, MI 49053
Dwarf fruit, other house plants.

BURNETT BROTHERS, INC.
92 Chambers Street, New York, NY 10007
Freesias and other bulbous plants.

W. ATLEE BURPEE COMPANY
Philadelphia, PA 19132; Clinton, IA 52733; and Riverside, CA 92502
Seeds, bulbs, supplies.

COOK'S GERANIUM NURSERY
712 North Grand, Lyons, KS 67544
Geraniums. Catalog 50¢.

CONARD PYLE STAR ROSES
West Grove, PA 19390
Miniature roses, clematis.

DE GIORGI COMPANY, INC.
Council Bluffs, IA 51501
Seeds and bulbs.

P. DE JAGER AND SONS, INC.
188 Asbury Street, South Hamilton, MA 01982
Bulbs for forcing.

FARMER SEED AND NURSERY COMPANY
Faribault, MN 55021
Dwarf citrus and other house plants.

FENNELL ORCHID COMPANY
26715 S.W. 157 Avenue, Homestead, FL 33030
Orchids.

HENRY FIELD SEED AND NURSERY COMPANY
407 Sycamore Street, Shenandoah, IA 51602
House plants and supplies.

FISCHER GREENHOUSES
Linwood, NJ
African violets and other gesneriads: supplies for growing house plants.

J. HOWARD FRENCH
Baltimore Pike, Lima, PA 19060
Bulbs for forcing.

GREEN ACRES NURSERY
14451 N.E. Second Street, North Miami, FL 33161
Palms.

HAUSERMANN'S ORCHIDS
2 N. 134 Addison, Villa Park, IL 60181
Unusual species orchids.

HILLTOP HERB FARM
Route 3, Box 216, Cleveland, TX 77327

S. M. HOWARD ORCHIDS
Seattle Heights, WA 98063
Unusual orchids.

JONES AND SCULLY, THE ORCHID PEOPLE
N.W. 24th Street Road @ 33rd Avenue
Miami, FL 33142
Orchids and other tropicals.

MICHAEL J. KARTUZ
92 Chestnut Street, Wilmington, MA 01887
House plants, many bred specially for fluorescent-light culture. Catalog 50¢.

LOGEE'S GREENHOUSES
55 North Street, Danielson, CT 06239
All kinds of house plants including rare and unusual exotics. Catalog $1.00.

LYNDON LYON
14 Mutchler Street, Dolgeville, NY 13329
African violets, gesneriads.

ROD McCLELLAN COMPANY
1450 El Camino Real,
South San Francisco, CA 94080
Orchids and supplies for growing.

MERRY GARDENS
Camden, ME 04843
Complete selection of house plants and herbs. Catalog $1.00.

MINI-ROSES
Box 4255 Station A, Dallas, TX 75208
Miniature roses.

NIES NUSERIES
5710 S.W. 37th Street, West Hollywood, FL 33023
Palms.

NUCCIO'S NURSERIES
3555 Chaney Trail, Altadena, CA 91002
Camellias.

GEORGE W. PARK SEED CO., INC.
Greenwood, SC 29647
All kinds of house plants, seeds, bulbs, fluorescent-lighting equipment, other supplies.

REASONER'S TROPICAL NURSERIES, INC.
4610 - 14th Street, West Bradenton, FL 33507

RIVERMONT ORCHIDS
P.O. Box 67, Signal Mountain, TN
Orchids.

SEQUOIA NURSERY
2519 East Noble Avenue, Visalia, CA 93277
Miniature roses, including basket and moss types.

SPIDELL'S FINE PLANTS
Junction City, OR 97448
African violets and other gesneriads.

FRED A. STEWART, INC.
1212 East Las Tunas Drive, San Gabriel, CA 91778
Orchids and supplies for growing.

SUNNYBROOK FARMS
9448 Mayfield Road, Chesterland, OH 44026
Herbs and house plants.

TINARI GREENHOUSES
2325 Valley Road, Huntington Valley, PA 19006
African violets, related plants, supplies and equipment.

WILSON BROTHERS
Roachdale, IN 46172
Geraniums and other house plants.

YOSHIMURA BONSAI COMPANY, INC.
P.O. Box 265, Briarcliff, NY 10501
Bonsai and supplies.

RUDOLPH ZIESENHENNE
1130 N. Milpas Street, Santa Barbara, CA 93003
Begonias.